National 5 Chemistry

How to Pass

SECOND EDITION

NATIONAL 5

Chemistry

Barry McBride

The Publishers would like to thank the following for permission to reproduce copyright material.

Photo credits

p.18 © denis cristo / 123RF; **p.48 top** © Barry McBride, **middle** © Kelpfish – Fotolia, **bottom** © Jack Star/Photodisc/Getty Images/ Environmental Concerns 31; **p.60 left** © Hallgerd – Fotolia, **right** © J Marshall - Tribaleye Images / Alamy Stock Photo, **p.63** © Jim Parkin / Alamy Stock Photo; **p.89** © Chris Mattison / Alamy Stock Photo.

Every effort has been made to trace all copyright holders, but if any have been inadvertently overlooked, the Publishers will be pleased to make the necessary arrangements at the first opportunity.

Orders: please contact Bookpoint Ltd, 130 Park Drive, Milton Park, Abingdon, Oxon OX14 4SE. Telephone: (44) 01235 827720. Fax: (44) 01235 400454 Lines are open from 9.00–5.00, Monday to Saturday, with a 24-hour message answering service. Visit our website at www.hoddereducation.co.uk. Hodder Gibson can also be contacted at hoddergibson@hodder.co.uk

First published in 2018 by
Hodder Gibson, an imprint of Hodder Education
An Hachette UK Company
211 St Vincent Street
Glasgow, G2 5QY

Impression number 5 4 3 2 1

Year 2022 2021 2020 2019 2018

Cover photo © Maksym Yemelyanov/stock.addoe.com
Illustrations by Aptara, Inc.
Typeset in 13/15 Cronos Pro (Light)/ Aptara, Inc.
Printed in Spain
A catalogue record for this title is available from the British Library.
ISBN: 978 1 5104 2086 1

Contents

Introduction

Students

This easy-to-read book delivers the National 5 Chemistry course in an enjoyable way. Read carefully and at a steady pace to give yourself time to understand the concepts and theories. Throughout the book there are Hints & tips to help you study and Key points that you must learn; pay particular attention to these, as they will help greatly. Key words are shown in bold and definitions are given at the end of the book.

At the end of each chapter, there are exam-type questions that will test your knowledge. These questions are written in a similar style to those you will see in your exam. They use the same language and are at the correct level, so it is important to practise answering all of the questions in full.

Try a question and then refer to the notes if you find it difficult. Answers to all the questions are provided at the end of the book.

Remember that the rewards for passing National 5 Chemistry are well worth it! Your pass will help you towards the future you want for yourself. In the exam, be confident in your own ability. If you are not sure how to answer a question, trust your instincts and just give it a go anyway. Keep calm, don't panic and **good luck!**

Teachers/lecturers and parents/carers

This book has been developed to cover the mandatory content as outlined in the National 5 Chemistry Course Support Notes. It also contains some National 4 content to refresh student's knowledge, before progressing further into the topic. It is written to be accessible to students, delivering concepts and theories in a 'straight-to-the point' manner, using clear and simple language. It is based on proven classroom methods of delivering difficult concepts in a style that the students can relate to and understand. This book has been designed both as a classroom aid and for a student's own personal study.

Exam details

To achieve a pass in National 5 Chemistry you must be successful in the two main components.

Component 1 – the assignment

You are required to submit an assignment that is worth 20% (20 marks) of your final grade. This assignment will be based on research and must include an experiment. It requires you to apply skills, knowledge and understanding to investigate a relevant topic in chemistry. Your school or college will provide you with the SQA's Candidate Guide. This gives guidance on what is required to complete the report and gain as many marks as possible.

Your assignment report will be marked by the SQA.

Component 2 – the question paper

The question paper will assess breadth and depth of knowledge and understanding in all of the three areas of National 5 Chemistry, and has a maximum score of 100 marks.

The question paper requires you to:

- make statements, provide explanations, and describe information to demonstrate knowledge and understanding
- apply knowledge and understanding to new situations to solve problems
- plan and design experiments
- present information in various forms, such as graphs and tables
- perform calculations based on information that is provided
- give predictions or make generalisations based on information that is provided
- draw conclusions based on information that is provided
- suggest improvements to experiments to make the results more accurate or to address safety concerns.

To achieve a 'C' grade in National 5 Chemistry you must achieve at least 50% of the 100 marks available when the two components (the question paper and the assignment) are combined. For a 'B' grade you will need 60%, while for an 'A' grade you must ensure that you gain as many of the marks available as possible and at least 70%.

The majority of the marks will be awarded for demonstrating and applying knowledge and understanding of the mandatory content of the course. The other marks will be awarded for applying scientific inquiry skills. Marks will be distributed equally across the three areas.

Open-ended questions will also be included in the exam.

Section 1 Chemical changes and structure

Chapter 1.1 Rates of reaction

Chemical reactions occur at different rates. They can be very fast, such as the reactions that cause explosions, or very slow, such as the corrosion of iron. In this chapter we will look at the factors that affect the rate of a chemical reaction and how the rate of a reaction can be calculated.

There are four factors that affect the **rate** of a chemical reaction:

- **temperature**
- **concentration**
- **surface area**
- the presence of a **catalyst**.

To understand how these factors affect the rate of a reaction we must first think about why chemical reactions take place. For a reaction to happen, reactant particles must collide with enough energy to allow them to combine to form new products.

Figure 1.1 Particles reacting

Temperature

Increasing the temperature of a reaction mixture increases the energy of the particles, making them move faster. This results in more successful collisions, which produces a faster chemical reaction.

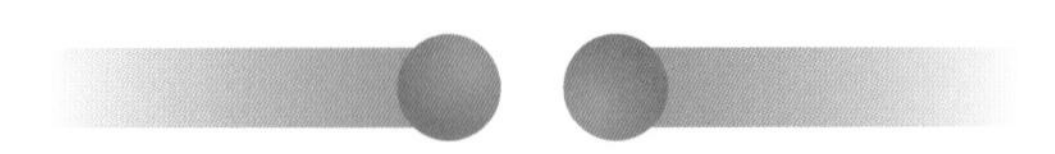

Figure 1.2 At higher temperatures fast-moving particles collide more often

> **Key points !**
>
> * The higher the temperature, the faster the reaction.
> * The lower the temperature, the slower the reaction.

Concentration

If the concentration of a substance increases, there are more particles present. This results in more collisions taking place, which produces a faster chemical reaction.

Figure 1.3 Increasing the concentration of particles increases the rate of a chemical reaction

Key points !

* The higher the concentration, the faster the reaction.
* The lower the concentration, the slower the reaction.

Surface area

When a substance is broken down into smaller pieces, its surface area is greatly increased. For example, 3 g of sugar powder has a much larger surface area than a 3 g sugar cube.

This increase in surface area results in a much faster reaction because more collisions can occur.

Figure 1.4 Powdered sugar has a greater surface area than cube sugar

Key points !

* The smaller the particle size, the greater the surface area, resulting in faster reactions.
* The larger the particle size, the smaller the surface area, resulting in slower reactions.

Catalysts

Catalysts are used a lot in industry as they allow chemical reactions to take place at lower temperatures, which saves energy and money. They also increase the rate of a chemical reaction.

Platinum and other transition metals are used in car exhaust systems to convert harmful gases into less harmful substances.

When a catalyst is added to a reaction mixture, it is *not* used up in the reaction. This means that all the catalyst added to the mixture can be recovered unchanged when the reaction is complete.

Key points !

* Catalysts speed up the rate of a chemical reaction but are not used up in the reaction, meaning they can be recovered when the reaction is complete.
* They are used in industry to save energy and money as they allow reactions to take place at lower temperatures.

Remember

Variable	Effect on reaction rate	Common examples
temperature	low temperature = slow reaction high temperature = fast reaction	If food is kept in the fridge, the reactions that cause the food to go off are slowed down, allowing the food to last longer.
concentration	low concentration = slow reaction high concentration = fast reaction	Chalk reacts much faster with a more concentrated acid.
surface area	large surface area = fast reaction small surface area = slow reaction	Chips cook much faster than whole potatoes, because chips have a larger surface area.
catalyst	catalysts speed up the rate of a reaction but are not used in the reaction	In the Haber process (see page 82) an iron catalyst is used to allow the production of ammonia to take place at a lower temperature.

Following the course of a reation

In chemistry, different measurements can be taken to follow the course of a reaction.

Two of the most common ways of monitoring the course of a reaction are by measuring the volume of gas produced (Figures 1.5 and 1.7) or measuring the loss in mass over a period of time (Figure 1.9).

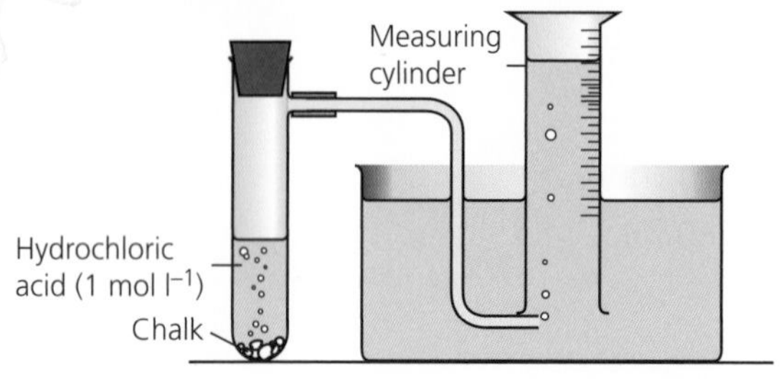

Figure 1.5 This experiment was set up to investigate the rate of reaction between chalk and hydrochloric acid

Figure 1.6 The carbon dioxide produced in this reaction can be collected by the upward displacement of air, because carbon dioxide gas is more dense than air

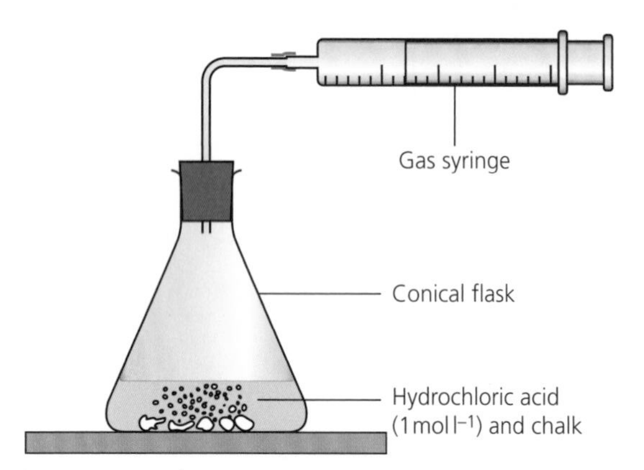

Figure 1.7 An alternative way to measure gas production

Hints & tips

The method shown in Figure 1.6 cannot be used to measure the volume of carbon dioxide produced because the gas is colourless.

The equipment shown in Figure 1.5 cannot be used to collect a soluble gas as it would dissolve in water. The method shown in Figure 1.7 is used when a soluble gas is produced by a reaction.

The measurements obtained from these reactions can be used to produce rate graphs that tell us many things about a chemical reaction.

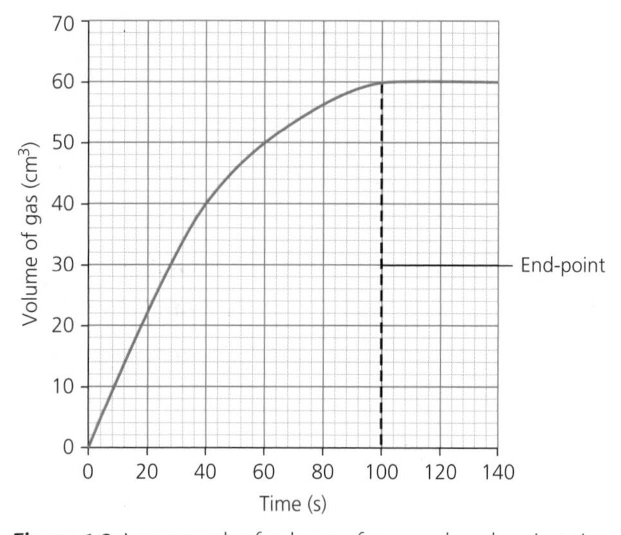

Figure 1.8 A rate graph of volume of gas produced against time

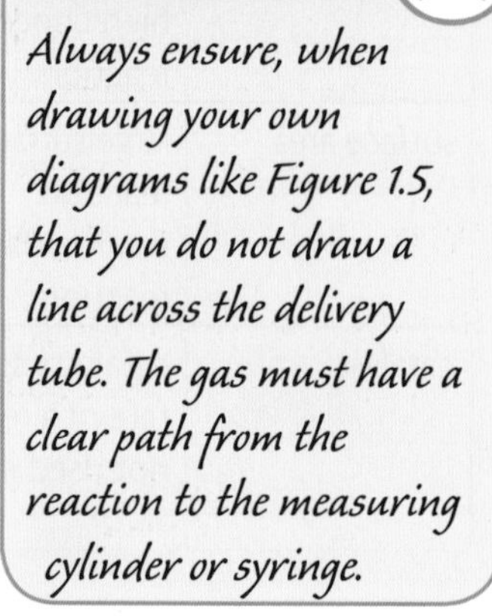

The point on the graph where the line flattens out (becomes horizontal) is called the **end-point** (shown by the dotted line). At this point, no gas is being produced and, therefore, the reaction has stopped.

Measuring the change in mass during a reaction can also give results to plot a rate graph.

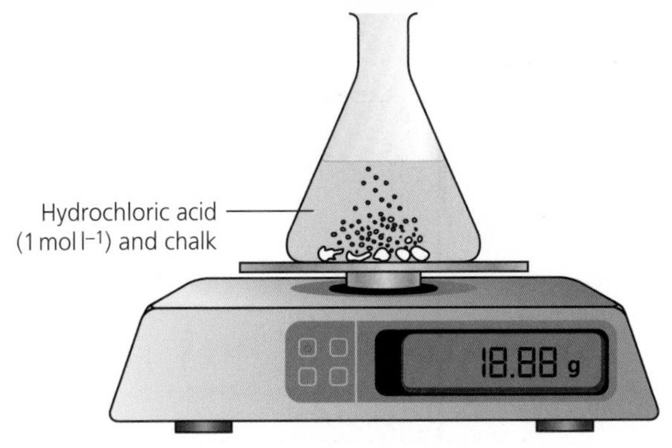

Figure 1.9 Measuring the change in mass during an experiment

In the experiment shown in Figure 1.9, the mass decreases as the reaction proceeds. This is because a gas is given off. The gas leaves the conical flask, resulting in a decrease in mass. The decrease in mass is clearly shown when a graph of the results is plotted (Figure 1.10).

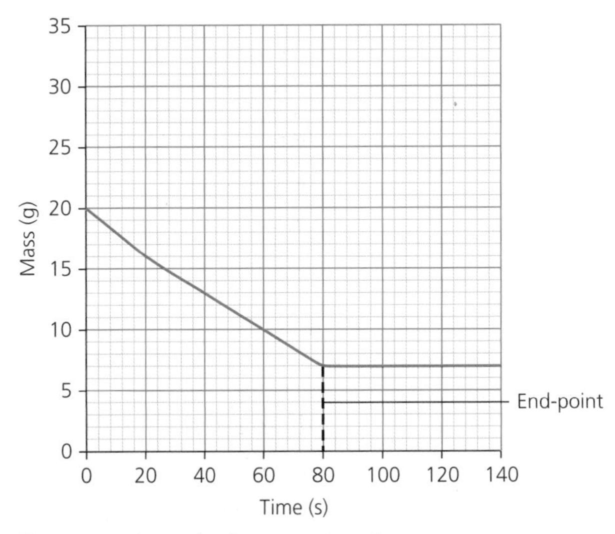

Figure 1.10 A graph of mass against time

Results tables and rate graphs can be used to calculate the average rate of a reaction using this equation:

$$\text{rate} = \frac{\Delta\text{quantity}}{\Delta\text{time}}$$

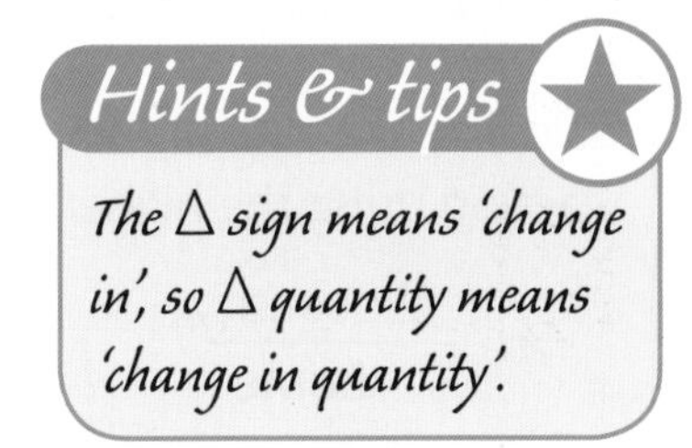

This equation can be found in page 3 of the Data Booklet. The units for rate depend on the units used to measure the quantity and the time. Examples of correct units are shown in the table.

Quantity unit	Time unit	Rate unit
cm^3	s	$cm^3\,s^{-1}$
g	min	$g\,min^{-1}$

Example 1

Calculate the rate of the reaction, in $cm^3 s^{-1}$, of the first 60 seconds of the reaction shown in Figure 1.8.

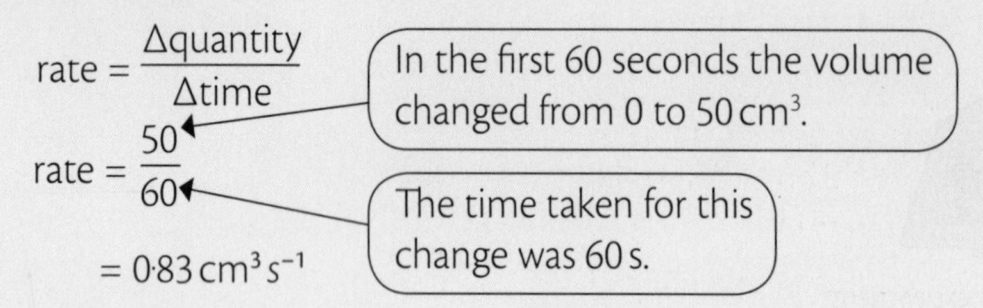

The experiment in Figure 1.8 was repeated using a higher concentration of acid and the results were drawn onto the same graph, using green ink.

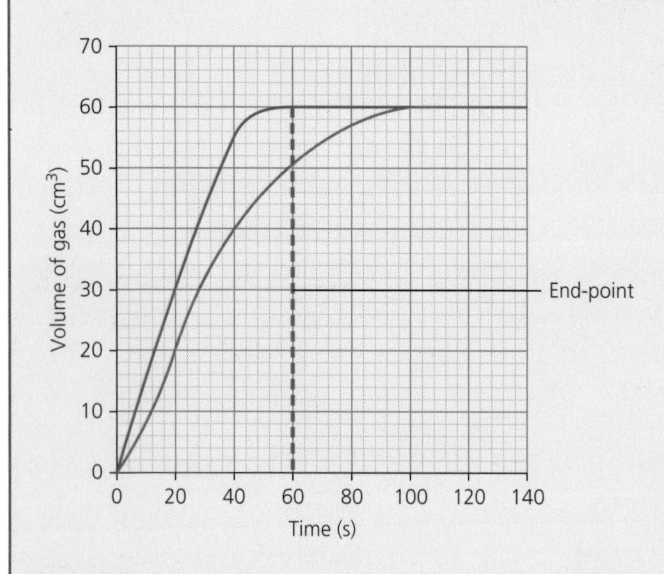

Figure 1.11 A graph of volume of gas produced against time for two different concentrations of acid

Calculate the rate of the reaction, in $cm^3 s^{-1}$, of the first 60 seconds of reaction using the higher concentration of acid (green line).

$$\text{rate} = \frac{\Delta\text{quantity}}{\Delta\text{time}}$$

In the first 60 seconds the volume changed from 0 to 60 cm^3.

$$\text{rate} = \frac{60}{60}$$

The time taken for this change was 60 s.

$$= 1\,cm^3\,s^{-1}$$

The rate of the second reaction is higher because the concentration of acid has been increased. This can also been seen in the graph, as the end-point of the second reaction is reached before the end-point of the first reaction.

Example 2 

The table shows the results obtained in an experiment to measure the change in mass during a reaction.

Time (s)	0	20	40	60	80	100
Mass (g)	100	80	70	67	65	65

Calculate the average rate of reaction between 20 and 60 seconds. Your answer must include the appropriate rate unit.

$$\text{rate} = \frac{\Delta\text{quantity}}{\Delta\text{time}}$$

Between 20 and 60 seconds the mass dropped from 80 to 67 = 13.

$$\text{rate} = \frac{13}{40}$$

The time taken for this change was 40 s.

$$= 0{\cdot}325\,\text{g}\,\text{s}^{-1}$$

Note the correct unit must be given in this answer. As the mass was in grams and the time in seconds the correct unit is $\text{g}\,\text{s}^{-1}$. Answers of $0{\cdot}3\,\text{g}\,\text{s}^{-1}$ and $0{\cdot}33\,\text{g}\,\text{s}^{-1}$ would also be correct.

Study questions 

1 A reaction between acid and chalk was carried out to monitor the rate of reaction. Which of the following changes would **not** affect the rate of the reaction?

a) change in size of test tube
b) change in particle size of chalk
c) change in concentration of acid
d) change in temperature of the acid

2 The rate of decomposition of hydrogen peroxide increased when the catalyst manganese dioxide was added. If 10 g manganese dioxide can be recovered after the reaction is complete, how much was present at the start?

a) 0 g
b) 5 g
c) 10 g
d) 20 g

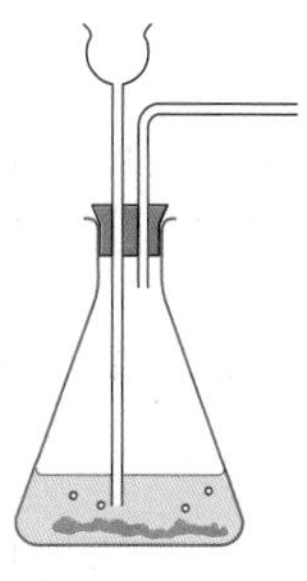

Figure 1.12

3 Joseph was investigating the rate of reaction between 3 g of chalk and an excess of $1\,\text{mol}\,\text{l}^{-1}$ hydrochloric acid. He monitored the reaction by measuring the volume of carbon dioxide gas produced over time.

a) Copy and complete Figure 1.12 to show how a sample of the gas could be collected.
b) Joseph recorded his results in a table. Draw a line graph to show these results.
c) Calculate the rate of reaction for the first 40 seconds. Your answer should include an appropriate rate unit.
d) Why does the rate of the reaction slow as time proceeds?
e) At what time did the reaction finish?
f) The reaction was repeated with a higher concentration of acid. Draw a line on the graph to represent the results that you would expect for this experiment.

Time (s)	Volume of CO_2 (cm^3)
0	0
10	10
20	20
30	28
40	35
50	41
60	45
70	48
80	50
90	51
100	51

4 In industry catalysts are often used to increase productivity. The catalysts used are usually in the form of a fine powder. Suggest why the catalysts are used in this form rather than as a lump.

5 The effectiveness of enzymes in catalysing the decomposition of hydrogen peroxide was investigated (Figure 1.13). A piece of liver was added to one test tube of hydrogen peroxide and the volume of oxygen gas produced was recorded. This experiment was repeated with a piece of potato.

The results obtained are shown on the graph in Figure 1.14.

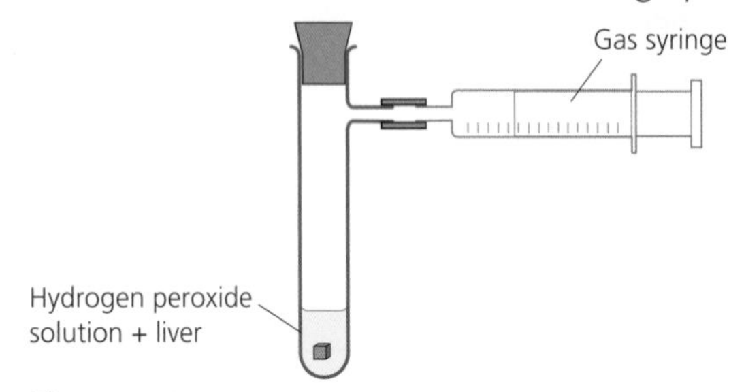

Figure 1.13

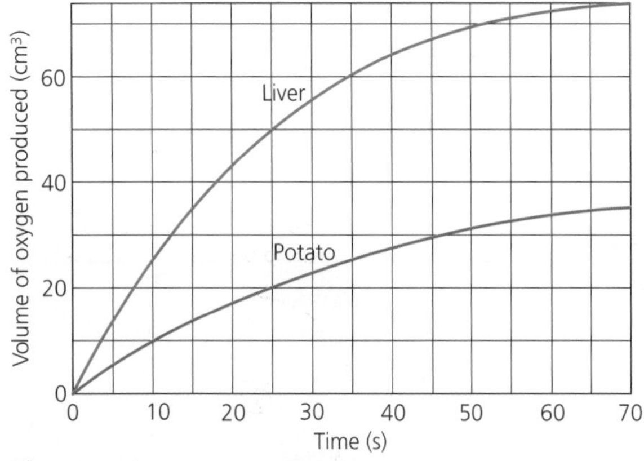

Figure 1.14

a) Calculate the rate of reaction, in $cm^3 s^{-1}$, for the two experiments over the first 25 seconds.

b) Which of the two, liver or potato, contained the most effective enzyme for this reaction?

c) What factors had to be kept constant to ensure the results are valid?

Chapter 1.2
Atomic theory

If we could look very closely at an element, we would see that an element is made of very tiny particles called atoms. In this chapter we will revise what makes up the atom and what holds atoms together.

The atom

To understand what holds **atoms** together, we need to look at the structure of the atom itself.

Atoms are made up of three smaller particles: electrons, protons and neutrons.

Key points

- **Electrons**: negatively charged particles that spin around the positive centre of the atom in circles called energy levels. Their mass is so small it is nearly zero.
- **Protons**: positively charged particles that are contained in the **nucleus** (the centre) of the atom. They have a mass of 1.
- **Neutrons**: neutrons are also contained in the nucleus of the atom but have no charge. They also have a mass of 1.

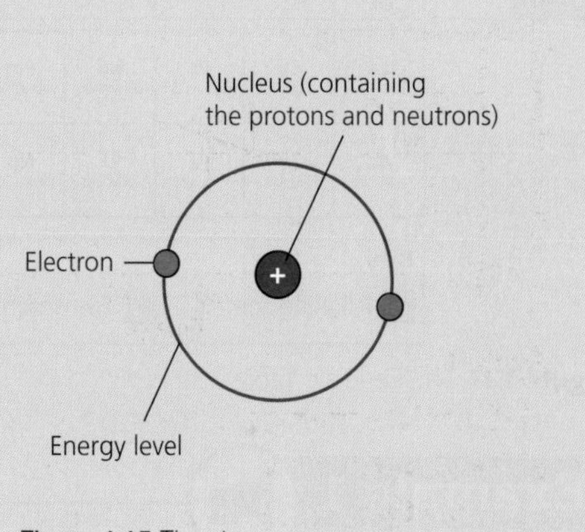

Figure 1.15 The atom

Atoms contain the same number of positive protons as negative electrons and, as a result, they are neutral; these particles with opposite charges cancel each other out. However, the nucleus at the centre of the atom is positively charged. This is because it contains all the positive protons.

Remember

Particle	Mass	Charge	Location
electron	approx. 0	negative	outside nucleus
proton	1	positive	inside nucleus
neutron	1	no charge	inside nucleus

Atomic number

Elements are arranged in the Periodic Table in order of increasing **atomic number**, with all the metals to the left of the table and the non-metals to the right. In this Periodic Table the atomic numbers are shown at the top of each element.

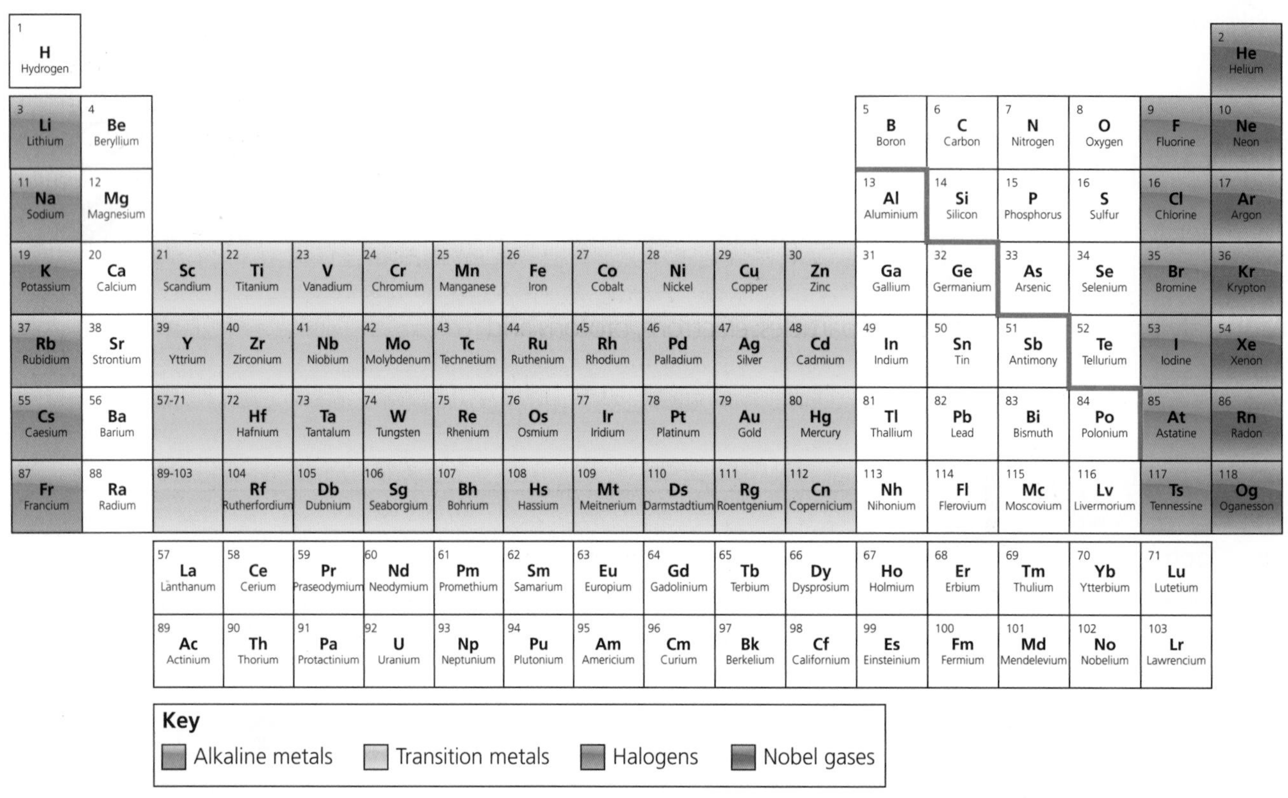

Figure 1.16 The Periodic Table

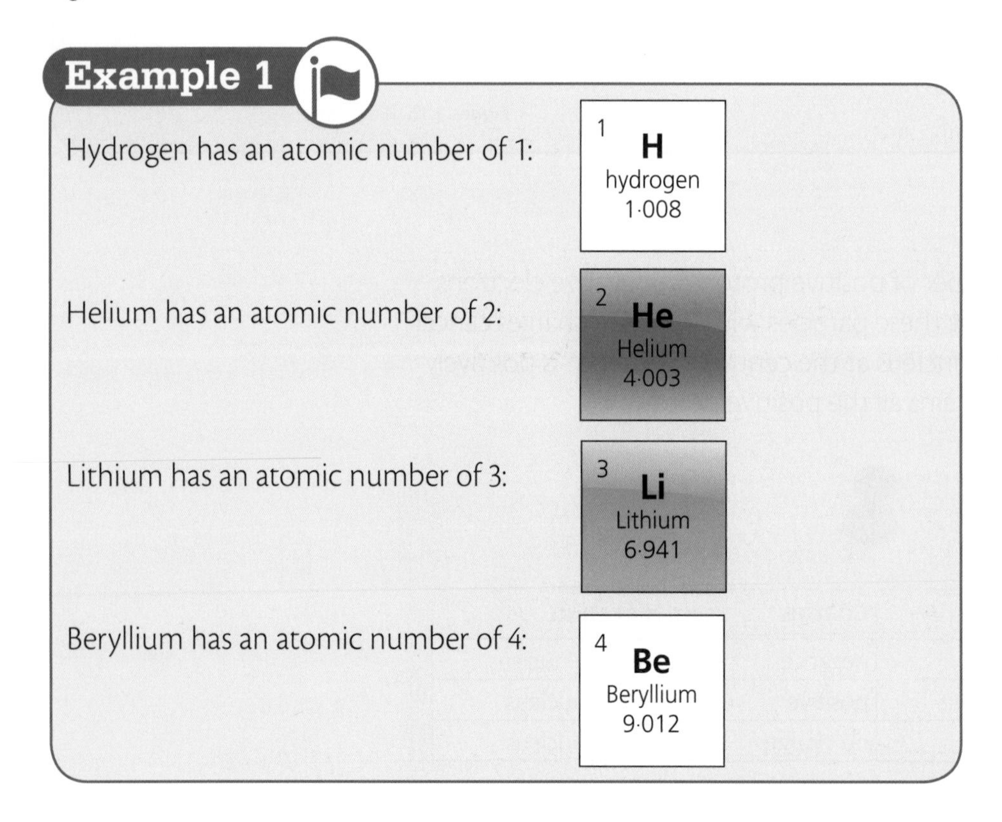

Example 1

Hydrogen has an atomic number of 1:

Helium has an atomic number of 2:

Lithium has an atomic number of 3:

Beryllium has an atomic number of 4:

Remember

Groups are columns in the Periodic Table. Li, Na and K are in the same group.

A period goes across the Periodic Table. Li, Be, B etc. are in the same period.

Key point

* The atomic number of an element tells you how many protons an atom of that element contains.

Example 2

The element magnesium has the atomic number of 12. This means that magnesium contains 12 protons and, because atoms are neutral, it must also have 12 electrons to cancel out the 12 protons.

You already know that the electrons in an atom are contained in energy levels (page 9). These energy levels can only hold a certain number of electrons.

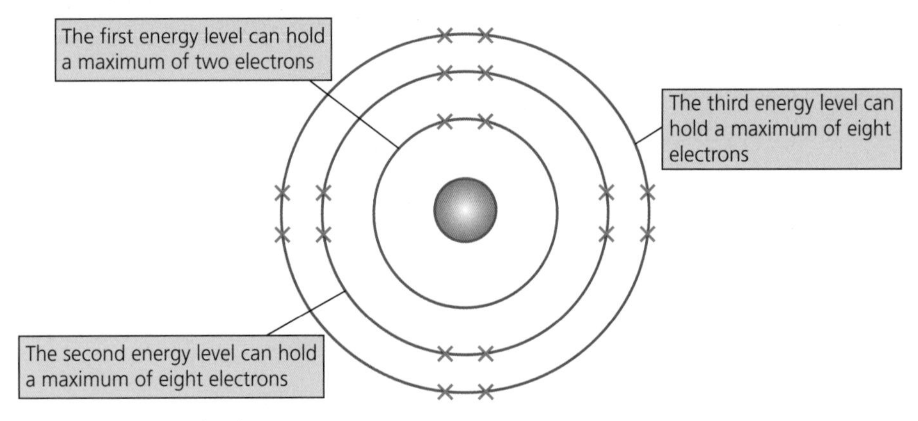

Figure 1.17 Energy levels in an atom

Now we can draw a diagram of an atom that shows how the electrons are arranged.

Example 3

Sodium has an atomic number of 11. This means that is has 11 protons and 11 electrons. The electrons can be placed in the energy levels as shown in Figure 1.18.

2 electrons would be placed in the first energy level,
8 in the second and
1 in the third

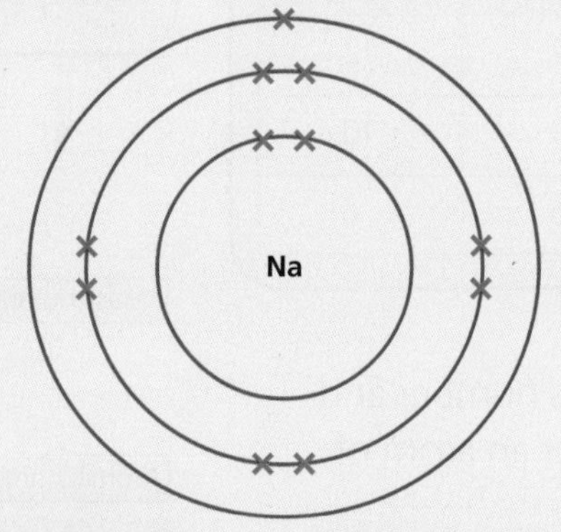

Figure 1.18 Electron energy levels in an atom of sodium

This gives sodium an electron arrangement of **2**, **8**, **1**.

The electron arrangements of all elements are given on page 6 of your Data Booklet.

The number of outer electrons in an element is very important, as it is these outer electrons that give elements their characteristic chemical properties. All elements in the same group of the Periodic Table have the same number of outer electrons – this is why they react in similar ways. All the **alkali metals** (group 1) have one outer electron and this makes them very reactive.

Hints & tips

For practice, draw the structure of each atom for the first 20 elements using the electron arrangements in your Data Booklet.

Mass number

We have seen how we can use the atomic number to calculate the number of protons and electrons in an atom. We can also work out how many neutrons there are in the nucleus of an atom.

To do this we look at the **mass number** of the atom.

Key point

* The mass number of an element is equal to the number of protons plus the number of neutrons.

Example 1

As we saw earlier, the atomic number of sodium is 11 (page 11). This tells us that sodium has 11 protons and, because it is neutral, it has 11 electrons.

Sodium also has a mass number of 23. This information allows us to calculate the number of neutrons contained in an atom of sodium.

total mass of sodium = 23

mass of protons = 11

mass of electrons = 0 (remember that electrons have no mass!)

mass of neutrons = 12 (23 – 11 protons = 12 neutrons)

Example 2

Some more examples are shown in the table.

Element	Atomic number	Mass number	Protons	Electrons	Neutrons
magnesium	12	24	12	12	12
fluorine	9	19	9	9	10
carbon	6	12	6	6	6

Hints & tips

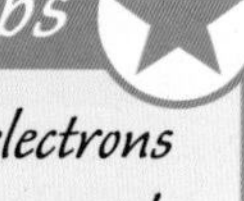

Remember that electrons have a mass of zero and, therefore, don't affect the mass of an element.

The nuclide notation of an element gives both the mass number and atomic number of the element. The nuclide notation for an atom of potassium is shown in Figure 1.19.

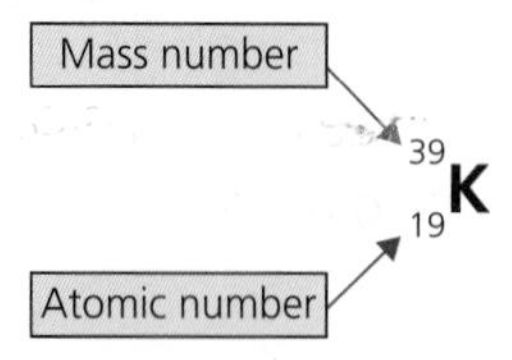

Figure 1.19 Nuclide notation for potassium

Nuclide notation allows us to calculate the number of protons, electrons and neutrons contained within the atom. Figure 1.20 gives the nuclide notation of potassium again but also shows what each piece of information means.

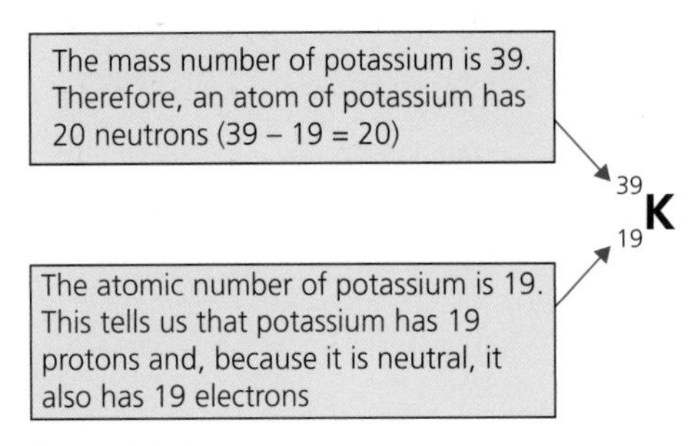

Figure 1.20

Hints & tips

The number of each particle type present in an atom can be worked out using your PEN:

- ✓ *Protons*
- ✓ *Electrons*
- ✓ *Neutrons!*

$^{27}_{13}Al$ **P** = 13 **E** = 13 **N** = 14 (27–13)

Figure 1.21 Remember how to use PEN!

Isotopes

The masses of the atoms of an element are not always the same. Just like people, atoms all have slightly different weights.

The atoms of an element always have the same number of protons (from the atomic number) but the number of neutrons can be different, resulting in a different mass number. These atoms with different numbers of neutrons are called **isotopes**.

Key point

* Isotopes are atoms with the same atomic number but different mass number.

The two carbon atoms shown in Figure 1.22 are isotopes as they have the same atomic number but they have different mass numbers.

$^{12}_{6}C$ $^{13}_{6}C$

Figure 1.22 Nuclide notation for two isotopes of carbon

When the nuclide notation is used to calculate the number of protons, electrons and neutrons in each atom, it can be seen that the difference in mass is due to the number of neutrons in each atom.

P = 6
E = 6
N = 6 (12 – 6) $^{12}_{6}C$

P = 6
E = 6
N = 7 (13 – 6) $^{13}_{6}C$

Figure 1.23 Remember how to use PEN!

Most elements have two or more isotopes, so the mass given in the Data Booklet is called the **relative atomic mass (RAM)**. This is an average mass of all the isotopes of a single element.

The relative atomic mass can be used to give an indication of the relative abundance of isotopes in a sample.

Example 1 

Copper has two isotopes:

$$^{63}_{29}Cu \text{ and } ^{65}_{29}Cu$$

The RAM of copper is 63·5.

The RAM indicates that copper-63 is the most abundant isotope due to the fact that the 63·5 is closer to copper-63 than to copper-65.

Example 2 

There is an equal abundance of both isotopes of magnesium and, as a result, the RAM lies in between the mass of each isotope. The RAM of magnesium is 24·5.

$$^{24}_{12}Mg \text{ and } ^{25}_{12}Mg$$

Ions

Atoms are neutral because they have an equal number of positive protons and negative electrons. But what happens when the numbers of protons and electrons are not equal?

When there is an imbalance of electrons to protons an **ion** is formed.

Key points 

- An ion is a charged particle that is formed when an atom *gains* or *loses* electrons.
- Metal atoms lose electrons to form positive ions.
- Non-metals gain electrons to form negative ions.

Noble gases (Group 8 elements) don't react and this is because their outer energy level is full of electrons. They have a stable electron arrangement.

Atoms of elements in other groups of the Periodic Table gain or lose electrons to achieve a stable electron arrangement (full outer energy level) like that of a noble gas.

Key point 

- The noble gases have a stable electron arrangement.

Example 1 

Neon, a noble gas, has eight outer electrons, whereas oxygen only has six outer electrons. To become stable like neon, oxygen gains two electrons from another atom.

Example 2

Metal: magnesium	Non-metal: chlorine
Magnesium has the electron arrangement of 2, 8, 2. Being a metal, Mg will lose two outer electrons to achieve the same electron arrangement as neon: 2, 8. $Mg \rightarrow Mg^{2+} + 2e^-$ This change creates an imbalance in the electron to proton ratio. $^{24}_{12}Mg^{2+}$ P = 12 E = 10 (12 – 2) N = 12	Chlorine has the electron arrangement of 2, 8, 7. Being a non-metal, Cl will gain one outer electron to achieve the same electron arrangement as argon: 2, 8, 8. $Cl + e^- \rightarrow Cl^-$ This change creates an imbalance in the electron to proton ratio. $^{35}_{17}Cl^-$ P = 17 E = 18 (17 + 1) N = 18

Figure 1.24 This wee joke tests to see how well you understand ions!

Study questions

1 Isotopes of an element have different
- **a)** mass numbers
- **b)** atomic numbers
- **c)** numbers of protons
- **d)** numbers of electrons.

2 An element has an atomic number of 11 and a mass number of 23. How many electrons are there in an atom of this element?
- **a)** 11
- **b)** 12
- **c)** 22
- **d)** 23

3 Which of the following electron arrangements is that of an element which has similar chemical properties to calcium?
- **a)** 2, 8, 1
- **b)** 2, 8, 2
- **c)** 2, 8, 3
- **d)** 2, 8, 4

4 Isotopes of the same element have identical
  a) numbers of neutrons
  b) atomic numbers
  c) mass numbers
  d) nuclei.

5 Which line in the table correctly describes a neutron?

	Mass	Charge
A	1	+1
B	1	0
C	0	−1
D	0	0

6 The table shows information about some particles. Which particle is a negative ion?

		Number of	
Particle	protons	neutrons	electrons
A	9	10	10
B	11	12	11
C	15	16	15
D	19	20	18

7 The nuclide notation can be used to calculate the number of protons, neutrons and electrons that an ion contains. The nuclide notation of a silver ion is shown.

$$^{107}_{47}Ag^{+}$$

  a) Copy and complete the table to show the number of each particle that this ion contains.

Particle	Number
Proton	
Electron	
Neutron	

  b) A sample of silver is found to contain two types of silver atom, ^{107}Ag and ^{109}Ag. What name is given to atoms of the same element with different masses?
  c) The relative atomic mass of silver is 108. What does this suggest about the relative abundance of these different atoms?

8 The nuclide notation shows the atomic number and mass number of an isotope. The nuclide notation for an isotope of neon is shown.

$$^{21}_{10}Ne$$

  a) An isotope of calcium has the atomic number of 20 and a mass number of 41. Write the nuclide notation for this isotope of calcium.
  b) How many protons and neutrons does this isotope of calcium contain?
  c) All the isotopes of calcium are electrically neutral. What does this suggest about the proton to electron ratio of each isotope?

Chapter 1.3
Bonding

We have revised what an atom is. Now we have to revise how atoms combine and what holds them together to form compounds. In this chapter we will revise the different types of bonds that form to hold atoms together.

Molecules

A **molecule** is two or more atoms joined together by **covalent bonds**. A molecule is usually made up of non-metal atoms only.

Water (H_2O) is an example of a molecule. It has two hydrogen atoms, combined to a single oxygen atom, by covalent bonds.

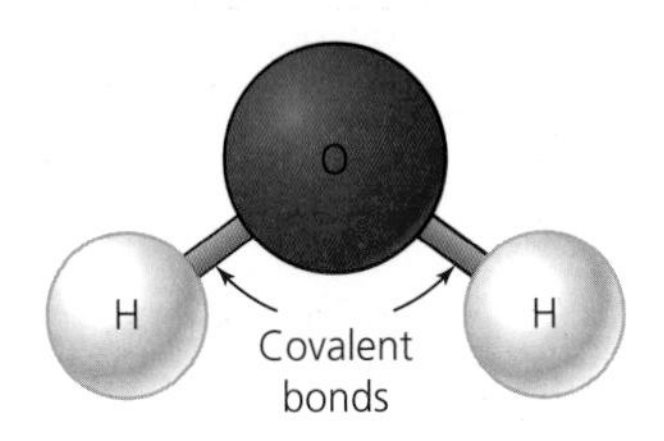

Figure 1.25 Model of a water molecule

Covalent bonds

All elements in the Periodic Table seek to become more stable, like a noble gas, and have a full outer energy level. For example, neon has eight outer electrons whereas oxygen only has six outer electrons. So to become stable like a noble gas, oxygen must gain two electrons. Hydrogen has only one outer electron and requires one more electron to achieve the same electron arrangement as helium.

Some element atoms can form bonds by sharing electrons with another atom. The bonds formed by sharing electrons in this way are called covalent bonds.

Hints & tips

The electron arrangements of elements are shown on page 6 of your Data Booklet.

Key points

- A molecule is two or more atoms held together by a covalent bond.
- A covalent bond is a shared pair of electrons between two non-metal atoms.

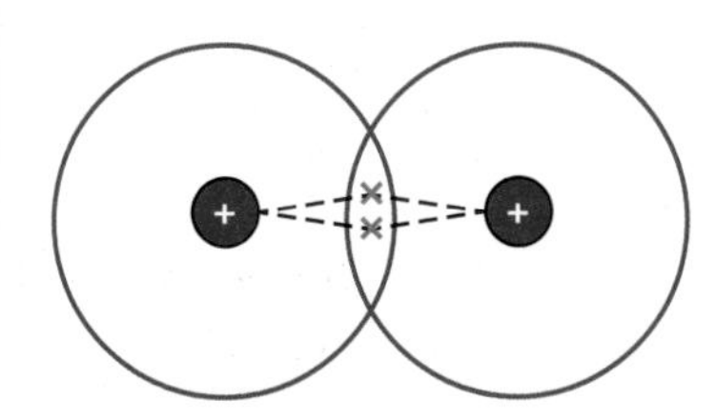

Figure 1.26 Electrostatic forces of attraction between nuclei and the shared electrons in a molecule

The atoms in a molecule are held together because of the electrostatic force of attraction between the positive nuclei of the atoms and the negatively charged electrons, as shown in Figure 1.26.

The saying 'opposites attract' can be used to describe how a covalent bond works. The positively charged nucleus of an atom is attracted to its negatively charged electrons. However, the nucleus of the other atom in the molecule is also attracted to them! This creates a 'tug-of-war' effect. Both nuclei try to pull the electrons toward themselves, creating a strong bond that holds the atoms together. This is shown by the dashed line in Figure 1.26.

Figure 1.26 is a molecule of hydrogen. A hydrogen atom has an electron arrangement of 1. So, by sharing an electron with another atom of hydrogen, both atoms have two outer electrons – giving them the same electron arrangement as the noble gas, helium.

Hints & tips

Remember the 'tug-of-war' description. It will help you to describe exactly how a covalent bond holds the atoms together.

Figure 1.27 How does this help us think about forces in a molecule?

Diatomic elements

Hydrogen is classed as a **diatomic element** as it exists, not as a single atom of hydrogen, but as a pair of hydrogen atoms that share electrons to become stable. There are other elements in the Periodic Table that are also diatomic. In other words, they exist as pairs in a molecule rather than as single atoms.

Example

Oxygen is a diatomic molecule; this is why its formula is O_2.

Key point

* A diatomic molecule is a molecule that contains two atoms only.

Remember

It is important to learn all the diatomic elements. These are listed in the table, along with a mnemonic to make it easier for you to learn them all. Or make up your own mnemonic.

Diatomic element	Symbol
fluorine	F_2
chlorine	Cl_2
oxygen	O_2
hydrogen	H_2
nitrogen	N_2
bromine	Br_2
iodine	I_2

→

Mnemonic
fancy
clancy
owes
him
nothing
but
ice

For all of the diatomic elements, you must be able to draw a diagram of the molecule showing all the outer electrons, like the hydrogen example in Figure 1.28. This shows the covalent bond formed by sharing electrons (the bond consists of an electron from one atom and an electron from the other atom).

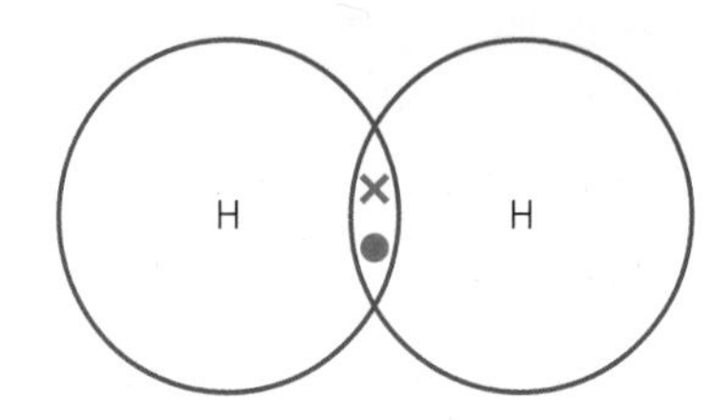

Figure 1.28 Bonding in hydrogen

Figure 1.29 shows a molecule of oxygen. Oxygen has the electron arrangement of 2, 6 and requires two electrons to become stable. Because of this, it shares two electrons with two electrons from another oxygen atom. This means that oxygen forms a double covalent bond.

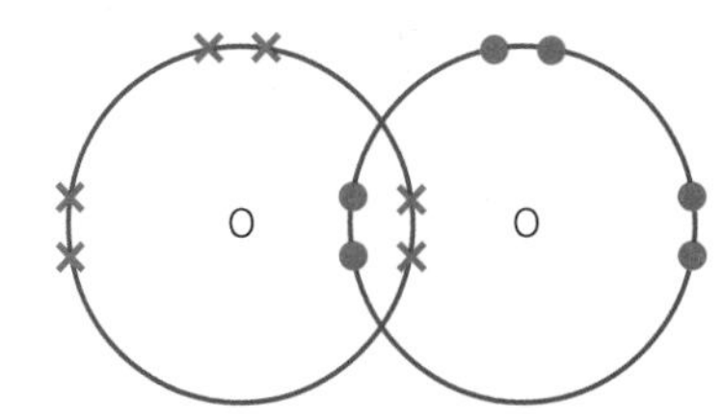

Figure 1.29 Bonding in oxygen

Hints & tips 

- ✓ *Draw a diagram showing all the electrons for all diatomic elements. This is a common task in questions in class tests and exams.*
- ✓ *To make sure you have drawn the diagrams correctly, just count the outer electrons. Each atom should have eight. Except for hydrogen – it should only have two. If they don't add up correctly, try again.*

Covalent compounds

So far we have looked at how the bonds are formed in molecules of elements, and the same rules apply when dealing with the bonding in covalent **compounds**.

Example 1

Methane (carbon hydride) has the chemical formula of CH_4. This formula occurs because the atoms of carbon and hydrogen share electrons to become stable.

An atom of carbon has four outer electrons and, therefore, requires a further four electrons to achieve a stable electron arrangement like a noble gas.

Hydrogen has only one electron and, therefore, requires only one more electron to achieve a stable electron arrangement.

For the two to combine and form a compound, in which both hydrogen and carbon have stable electron arrangements, the carbon atom requires four hydrogen atoms to supply four electrons.

Sharing electrons in this way allows both hydrogen and carbon to have stable electron arrangements and this is the reason why methane (carbon hydride) has the formula CH_4.

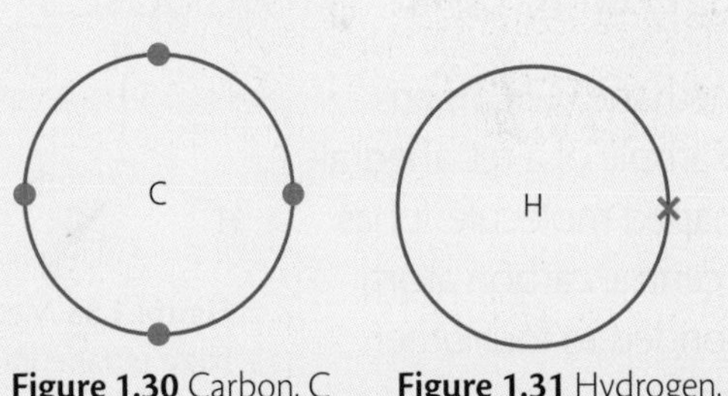

Figure 1.30 Carbon, C **Figure 1.31** Hydrogen, H

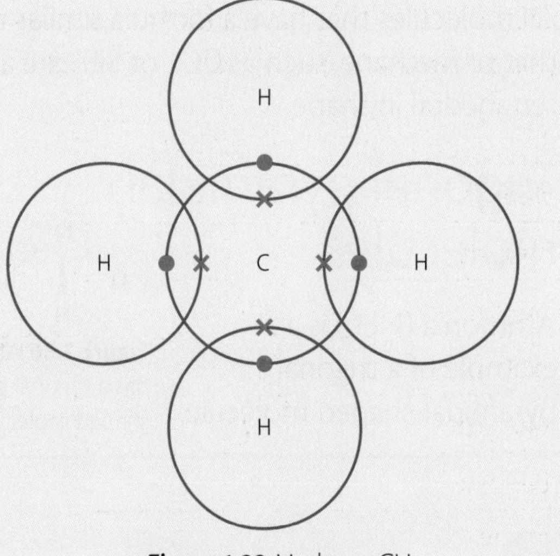

Figure 1.32 Methane, CH_4

Some other examples of covalent compounds are shown in Figures 1.33 and 1.34.

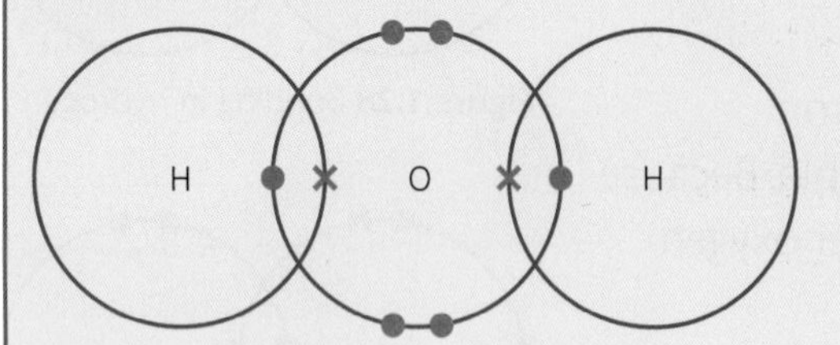

Figure 1.33 Water, H_2O

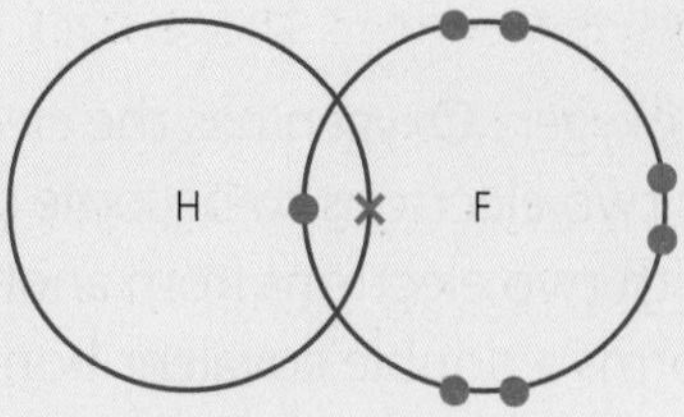

H F

Figure 1.34 Hydrogen fluoride, HF

Note that a molecule of hydrogen fluoride contains only two atoms and can, therefore, be classed as a **diatomic compound**.

Shapes of molecules

The bonds that are formed give molecules a distinctive shape.

The shapes are caused by the repulsion of electrons that are in the bonds.

Particles of the same charge will move away from each other. So the electrons in a covalent bond repel the electrons in other covalent bonds, causing them to move as far away from each other as possible. This creates molecules with different shapes.

Tetrahedral molecules

Methane (CH_4) is an example of a tetrahedral-shaped molecule. It has a central carbon atom bonded to four other atoms.

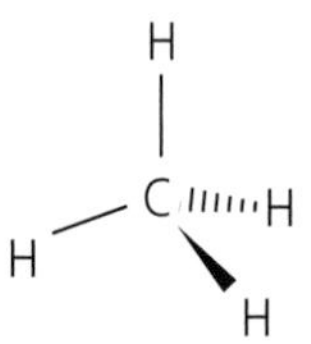

Figure 1.35 Methane has a tetrahedral-shaped molecule

All molecules that have a formula similar to that of methane, such as CCl_4 or SiF_4, are also tetrahedral in shape.

Trigonal pyramidal molecules

Ammonia (NH_3) is an example of a trigonal pyramidal-shaped molecule.

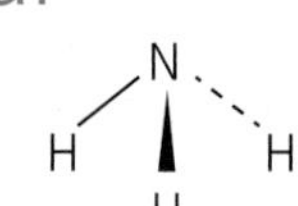

Figure 1.36 Ammonia has a trigonal pyramidal-shaped molecule

Angular molecules

Water (H_2O) is an example of an angular-shaped molecule.

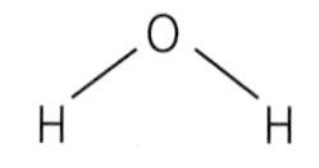

Figure 1.37 Water is an angular-shaped molecule

Linear molecules

Hydrogen fluoride (HF) is an example of a linear molecule.

H — F

Figure 1.38 Hydrogen fluoride has a linear molecule

Ionic bonding

In the previous chapter we looked at charged particles called ions, and the fact that metals form positive ions by losing electrons, and non-metals form negative ions by gaining electrons. (Look back at page 14 to refresh your memory before reading on.)

The saying 'opposites attract' can be used to describe an **ionic bond**. The attraction between the oppositely charged particles is called an electrostatic attraction.

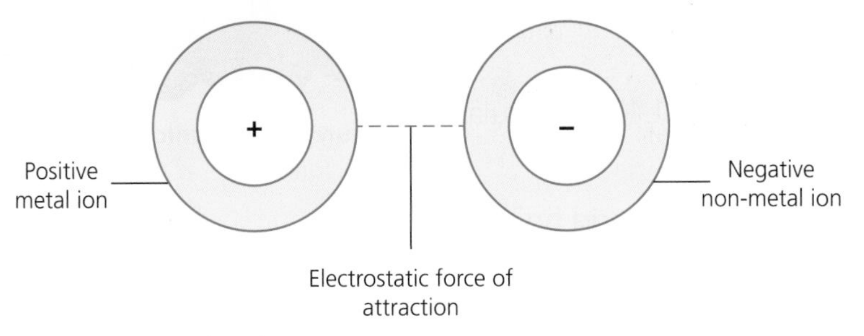

Figure 1.39 Ionic bond

Key point

* An ionic bond is an electrostatic force of attraction between a positive metal ion and a negative non-metal ion.

The type of bond formed between atoms has a major impact on the properties of that substance, so it is very important to be able to identify a substance as being covalent or ionic.

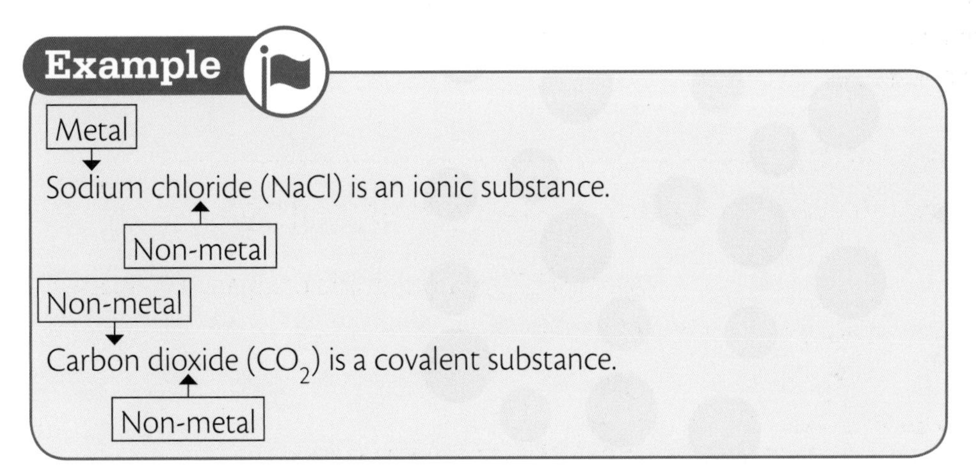

It is important to note that there are some exceptions to the general principles of

- metal + non-metal = ionic bonding
- non-metal + non-metal = covalent bonding

The properties of a substance, such as melting point and solubility, give us a much more accurate idea of its bonding. We will revise properties in the next chapter.

Chapter 1.4
Properties related to bonding

The properties of a substance depend on the types of bond that hold its atoms or ions together. In this chapter we will revise how bonding affects the properties of a substance.

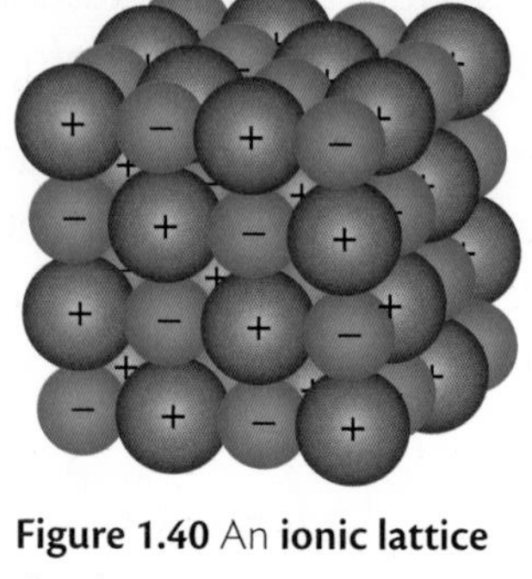

Figure 1.40 An **ionic lattice** structure

Ionic substances

Ionic compounds form an **ionic lattice** structure. Figure 1.40 is a regular arrangement of metal and non-metal ions.

This structure creates compounds with high melting points and boiling points, which also conduct electricity when molten or in solution but *never* when solid.

Ionic compounds dissolve easily in water. When they do this, the lattice breaks up completely to form free ions, allowing them to be surrounded by water molecules.

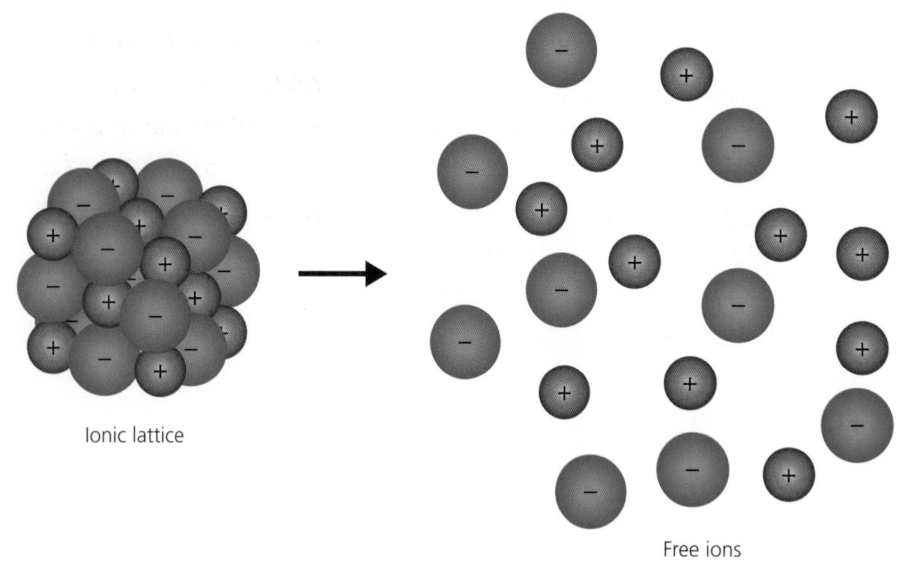

Figure 1.41 An ionic lattice dissolves to give free ions

When dissolved in water or molten, ionic compounds conduct electricity as the ions are free to carry the electrical charge.

The high melting point of ionic compounds is due to the millions of strong ionic bonds which must be broken, requiring lots of energy.

Key points

- Ionic compounds have high melting points and boiling points.
- Ionic compounds are soluble in water.
- Ionic compounds conduct electricity but only when molten or in solution because the ions are free to move.

Covalent substances

There are two types of covalent substance:

- discrete molecular substances
- covalent network substances.

Covalent network substances have a network of strong covalent bonds within one giant structure. As a result, they have very high melting points and are very hard. Diamond, for example, is a covalent network structure made of carbon atoms and is used in tools for cutting through rock.

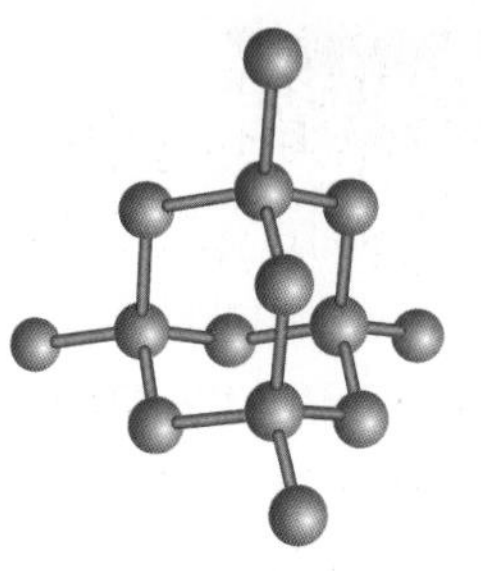

Figure 1.42 The covalent network structure of diamond

The many carbon atoms are held together by strong covalent bonds, which give diamond the properties of being very hard and having a very high melting point. Covalent network substances do not conduct electricity, with the exception of graphite. Graphite has delocalised (free-to-move) *electrons*, which allow it to conduct electricity.

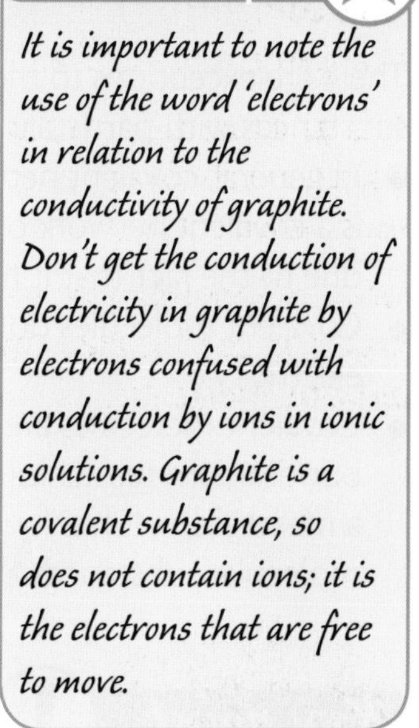

Hints & tips

It is important to note the use of the word 'electrons' in relation to the conductivity of graphite. Don't get the conduction of electricity in graphite by electrons confused with conduction by ions in ionic solutions. Graphite is a covalent substance, so does not contain ions; it is the electrons that are free to move.

Key points

* Covalent network substances have very high melting and boiling points.
* Covalent network substances are insoluble in water.
* Covalent network substances do not conduct electricity, with the exception of graphite because it has delocalised electrons.

Covalent molecules have strong covalent bonds within molecules but only weak attractions between molecules. This is demonstrated by substances such as oxygen and water, which have low melting and boiling points. This means that covalent molecules are liquids, gases or low melting-point solids at room temperature. Like covalent network compounds, they do not conduct electricity in any state, water being the only exception to this rule.

Key points

* Covalent molecules have strong covalent bonds within molecules, but only weak forces of attraction between molecules.
* Covalent molecules have low melting and boiling points.
* Covalent molecules do not conduct electricity, with the exception of water.

When a covalent molecular substance, such as ice, melts only the weak forces of attraction between the molecules, called intermolecular forces, are broken; not the strong covalent bonds.

Example

Ice, liquid water and steam are all made of water, proving that the bonds which make the water molecules are still intact after melting of the solid and **evaporation** of the liquid.

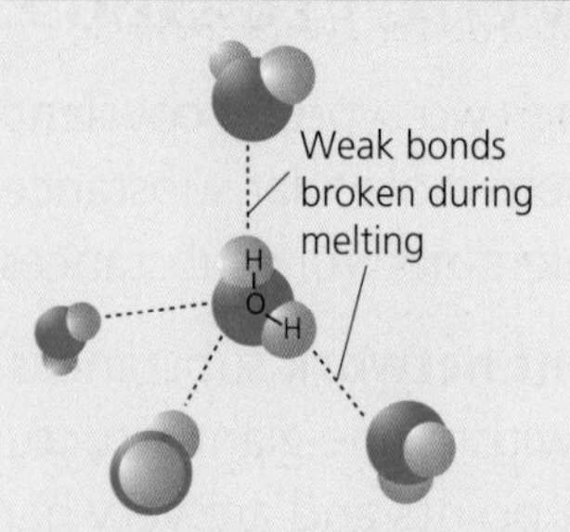

Figure 1.43 Weak bonds between molecules are broken when covalent molecular substances are melted

There are exceptions to these general rules about the properties of substances with particular types of bond:

- In general, covalent networks do not conduct electricity, *but* graphite is a covalent network of carbon atoms that does conduct electricity due to the fact that it has free-moving electrons.
- Covalent molecules do not conduct electricity, *but* water can conduct electricity.
- Covalent molecules are usually composed of non-metal atoms only, but titanium tetrachloride ($TiCl_4$) is a covalent molecule that contains a metal atom. It has low melting and boiling points like other covalent molecular substances.

Key point

* The table below is a summary of the information on properties of substances linked to bonding. It is very important to learn the table as it is covered in all exams.

Remember

Bonding and structure	Example	Melting point and boiling point	Conduction of electricity
covalent network	diamond	very high	non-conducting (except graphite)
ionic lattice	sodium chloride (salt)	high	only when molten or in solution (ions are free to move)
covalent molecule	carbon dioxide	low	non-conducting (except water)

Study questions ?

1 Which of the following substances exists as diatomic molecules?
  a) calcium oxide
  b) carbon dioxide
  c) carbon monoxide
  d) carbon tetrachloride

2 The diagram in Figure 1.44 is an example of
  a) an ionic lattice
  b) a metallic lattice
  c) a covalent network
  d) a covalent molecule.

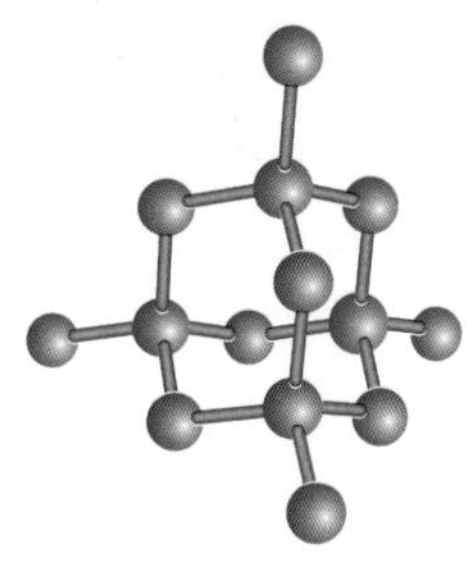

Figure 1.44

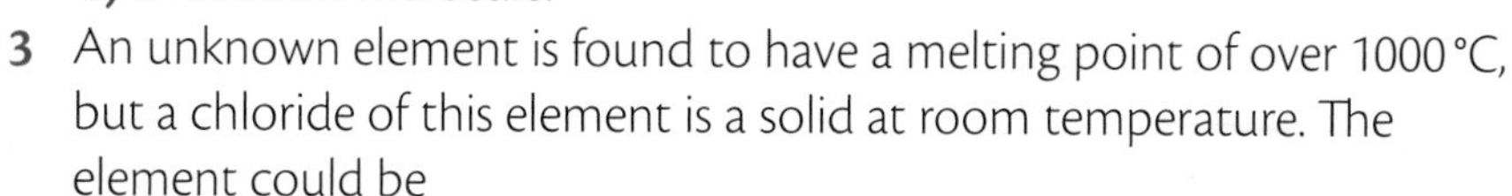

3 An unknown element is found to have a melting point of over 1000 °C, but a chloride of this element is a solid at room temperature. The element could be
  a) sulfur
  b) carbon
  c) copper
  d) hydrogen.

4 In a molecule of ammonia, NH_4, all the atoms are held together by strong covalent bonds.
  a) Explain fully how a covalent bond holds atoms together.
  b) i Draw a diagram of an ammonia molecule showing all the outer electrons.
     ii Draw a diagram to show the shape of an ammonia molecule.
  c) Water and methane are also covalent molecules.
     i Draw and name the shapes of both water and methane molecules.
     ii Water, methane and ammonia all have low boiling points. Explain fully why covalent molecular substances have low melting points.

5 The properties of four different substances are shown in the first table. Copy and complete the second table using the letters from the first table to show the type of bonding present in each substance.

Substance	Melting point (°C)	Boiling point (°C)	Electrical conduction
A	−77	−33	no
B	1883	2503	no
C	773	1407	when molten
D	1538	2862	yes

Substance	Bonding structure
	metallic lattice
	ionic lattice
	covalent network
	covalent molecular

6 Titanium tetrachloride ($TiCl_4$) has a melting point of −24 °C.
  a) Suggest the type of bonding in titanium chloride.
  b) Draw a diagram to show the shape of a titanium chloride molecule.

Chapter 1.5
Chemical formulae

H_2O, CO_2, NaCl – where do chemists get these formulae? In this chapter we will practise how to produce chemical formulae.

Naming compounds

The name of a compound comes from the elements that it contains. Naming compounds is fairly easy if you follow these rules:

1. Most compound names that end in '**–ide**' contain only the two elements mentioned in the name. For example:

   calcium + oxygen → calcium oxide

2. If the compound contains three or more elements, one of which is oxygen, then the compound name will end in '**–ate**' or '**–ite**'. For example:

   calcium + carbon + oxygen → calcium carbonate

Valency

To understand formulae, we must learn about **valency**.

Valency is the number of bonds that an element can form. The number of bonds an element can form is based on the number of outer electrons the atom contains. The valency of an element in groups 1 to 8 can be worked out using the Periodic Table, as shown in Figure 1.45.

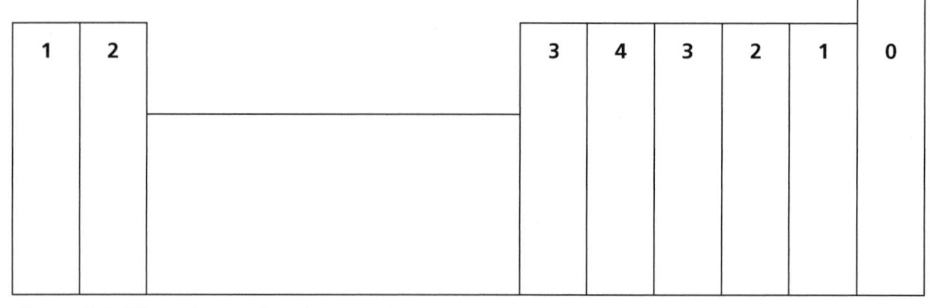

Figure 1.45 The valency of each group

This means, for example, that all the alkali metals have a valency of 1.

Formulae

The formulae of covalent molecular compounds give the number of atoms present in the molecule. In ionic and covalent network compounds the formulae give the simplest ratio of ions or atoms in the substance.

To write a chemical formula it is best to use the **S.V.S.D.F** system as shown in the examples below.

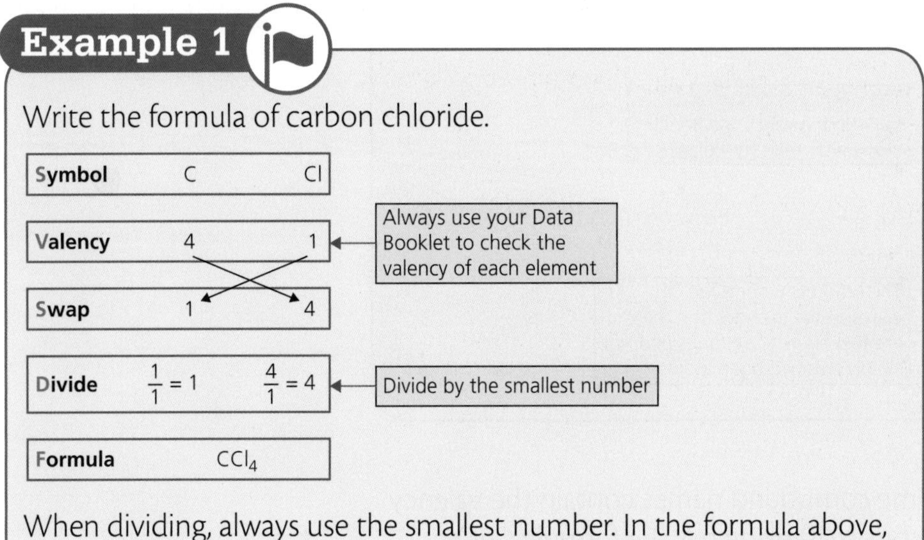

When dividing, always use the smallest number. In the formula above, the smallest number is 1, which makes no difference to the formula.

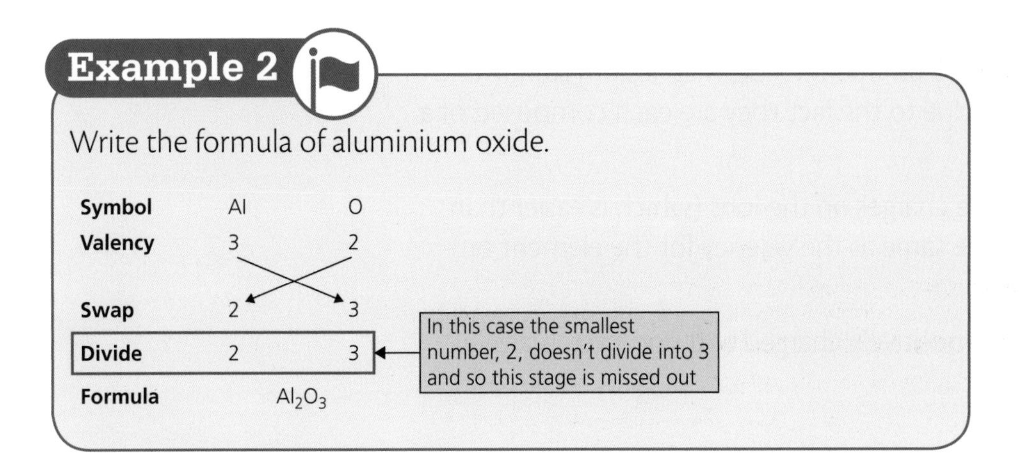

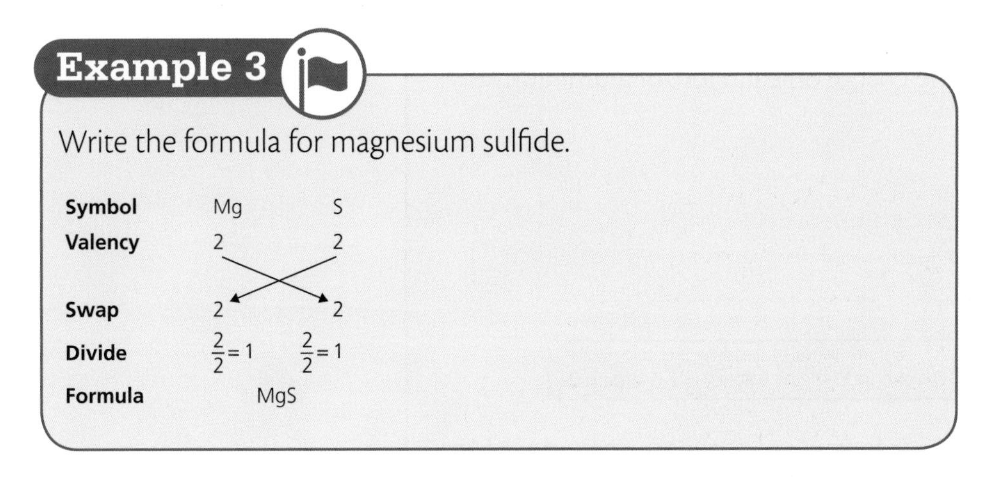

More complex formula involving group ions can also be produced using the same system.

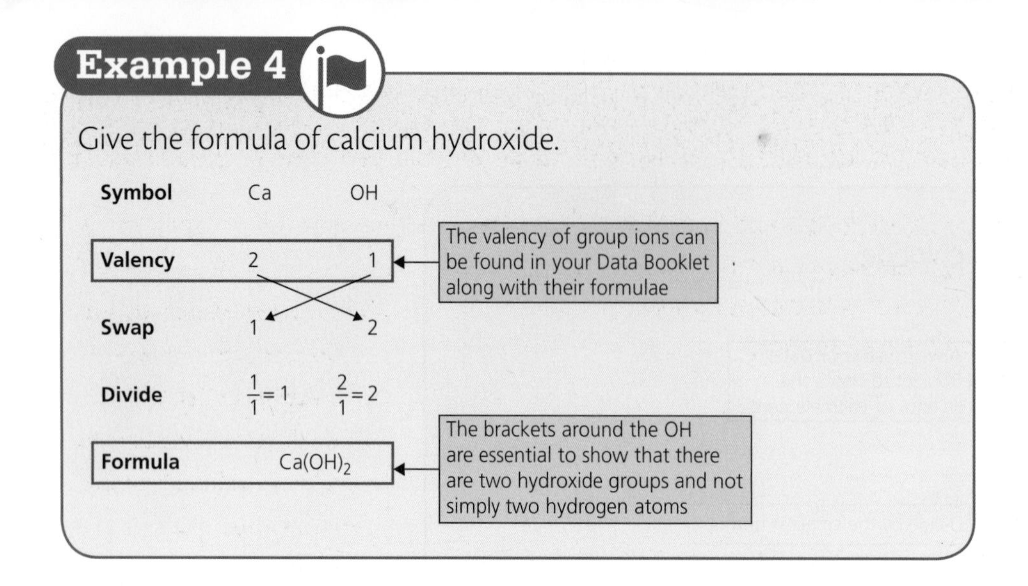

It is important to note that some compound names contain the valency of the metal ion, for example copper(II) oxide has the formula CuO because copper has the valency of 2, which is indicated by the roman numeral (II).

Formulae for ionic compounds can be taken a step further to form ionic formulae. Examples 2, 3, and 4 (aluminium oxide, magnesium sulfide and calcium hydroxide) are ionic, due to the fact they are each composed of a metal and a non-metal.

The ionic formula includes the charges on the ions (which is easier than it sounds!). The charges are the same as the valency for the element or group ion.

Remember that all metals are positively charged and non-metals are negatively charged.

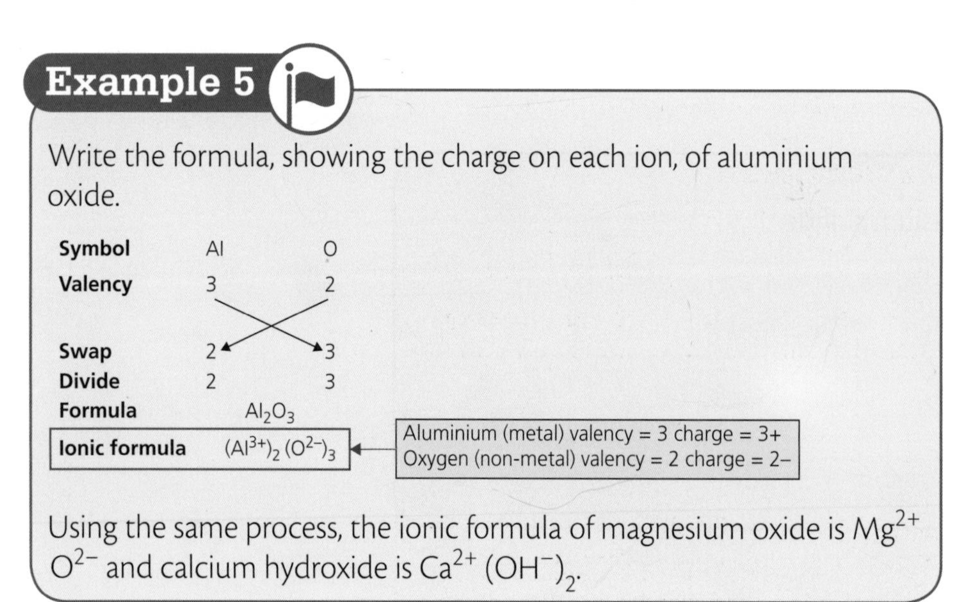

Chapter 1.6

The mole

When you think of a mole, do you think of a small furry animal that lives under the ground? Not anymore, because in chemistry one mole is equal to the gram formula mass of a substance. In this chapter we will revise mole calculations, including how to balance equations.

Hints & tips

The relative atomic masses of selected elements are listed on page 7 of the Data Booklet. These can be used to calculate the gram formula mass of a substance.

Calculating mass

The **gram formula mass** (**GFM**) of a substance is the mass of one **mole** of that substance.

Example 1 

Calculate the gram formula mass of one mole of potassium oxide K_2O.

Add all the relative atomic masses together, remembering to multiply the mass if there is more than one atom.

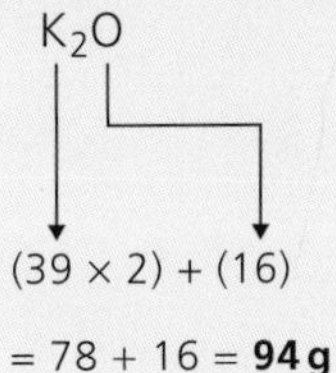

So the gram formula mass of potassium oxide is 94 g.

Example 2 

Calculate the gram formula mass of magnesium hydroxide, $Mg(OH)_2$.

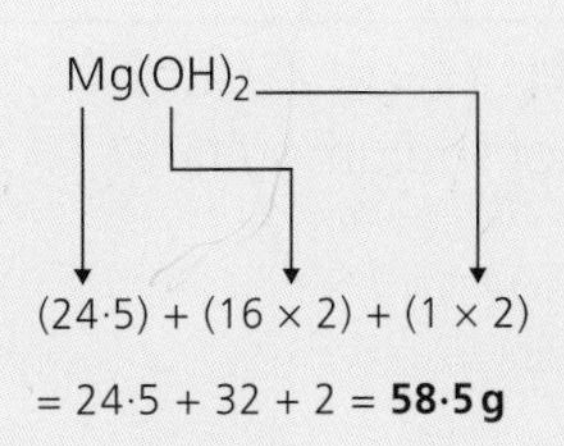

Example 3 

Calculate the gram formula mass of ammonium sulfate, $(NH_4)_2SO_4$.

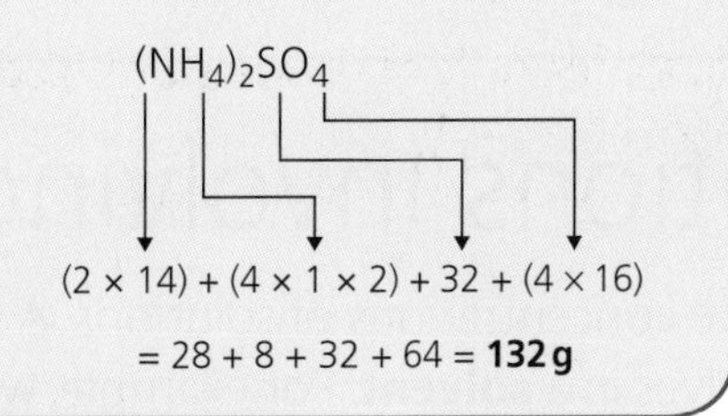

Mole calculations

Mole calculations can be done easily using the simple equation triangle in Figure 1.46. A version of this equation triangle is also on page 3 of the Data Booklet.

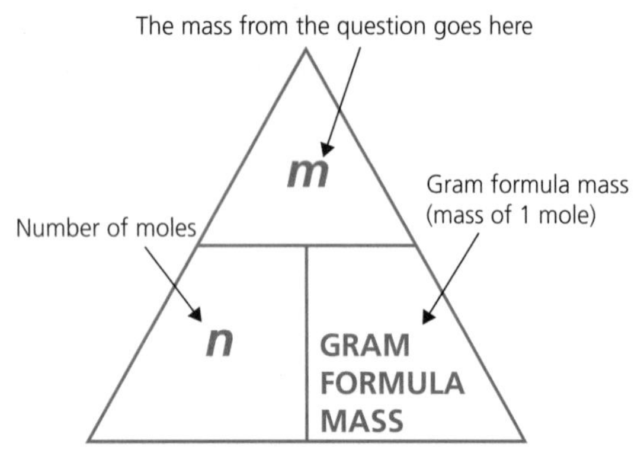

Figure1.46 Equation triangle for mole calculations

How many moles are present in 25 g of calcium carbonate?

Firstly, calculate the gram formula mass of calcium carbonate.

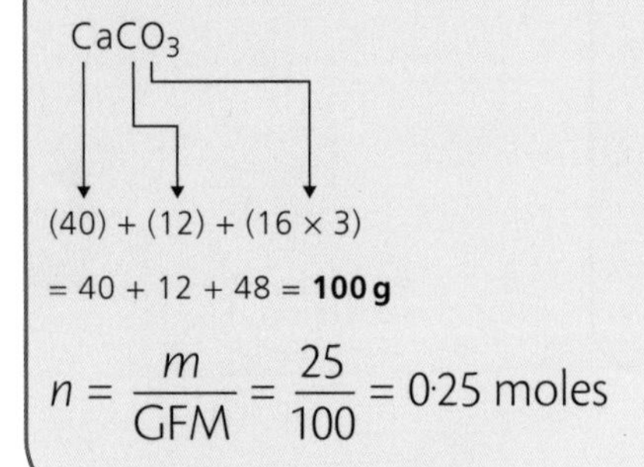

$$n = \frac{m}{GFM} = \frac{25}{100} = 0{\cdot}25 \text{ moles}$$

Example 2

What is the mass of 0·1 moles of sodium sulfate?

Na_2SO_4

(23 × 2) + 32 + (16 × 4)

= 46 + 32 + 64 = **142 g**

$m = n \times GFM = 0{\cdot}1 \times 142 = \mathbf{14{\cdot}2\,g}$

More calculations involving moles

Moles are also used to calculate concentrations of solutions. A **solution** is produced when a **solute** dissolves in a **solvent**. For example, when sodium chloride (salt) is added to water, a solution of salt water is formed. In this situation, the salt is the solute and the water is the solvent.

Remember

The solvent is where it went.
The solute is what you put.
The solution is what you're producin'!

To calculate the concentration of a solution, you need another equation triangle like the one in Figure 1.47. A version of this equation triangle is on page 3 of the Data Booklet.

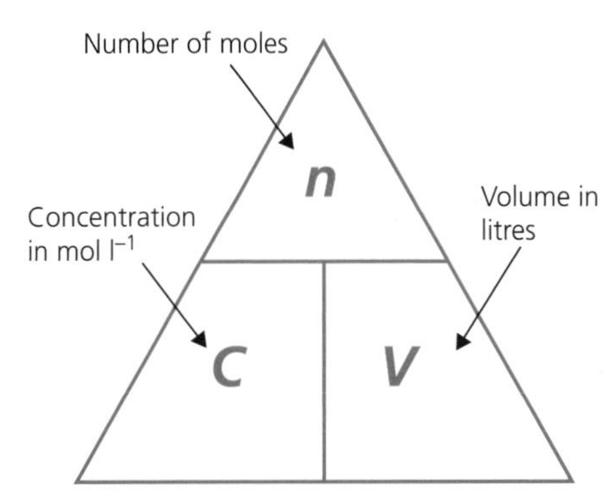

Figure 1.47 Equation triangle for concentration calculations

When using the equation triangle, the volume must be in litres. To convert cm^3 to litres divide by 1000. For example, if the volume is given as 250 cm^3, then in litres this is 0·25 l $\left(\frac{250}{1000} = 0{\cdot}25\right)$.

Example 1

Calculate the concentration of a sodium hydroxide solution if 2 moles are dissolved in 500 cm^3 of water.

n = 2 moles

$V = 500\,\text{cm}^3 = 0{\cdot}5\,\text{l}$

$C = \frac{n}{V} = \frac{2}{0{\cdot}5} = 4\,\text{mol l}^{-1}$

The calculations we have just looked at are commonly combined to produce one, more complicated calculation, which involves using both triangles at once!

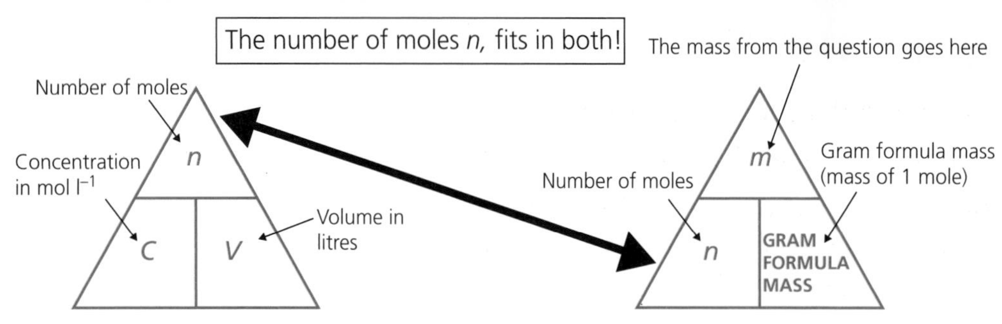

Figure 1.48 How the equation triangles fit together

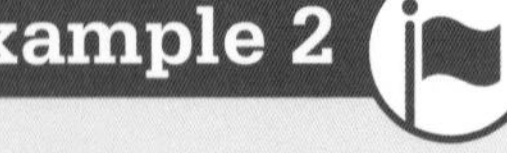

Example 2

Calculate the mass of potassium hydroxide required to make 250 cm^3 of 2 mol l^{-1} potassium hydroxide solution.

$V = 250\ cm^3 = 0{\cdot}25\ l$

$C = 2$

$n = C \times V = 2 \times 0{\cdot}25 =$ **0·5 moles**

This can be used to calculate the mass using the gram formula mass.

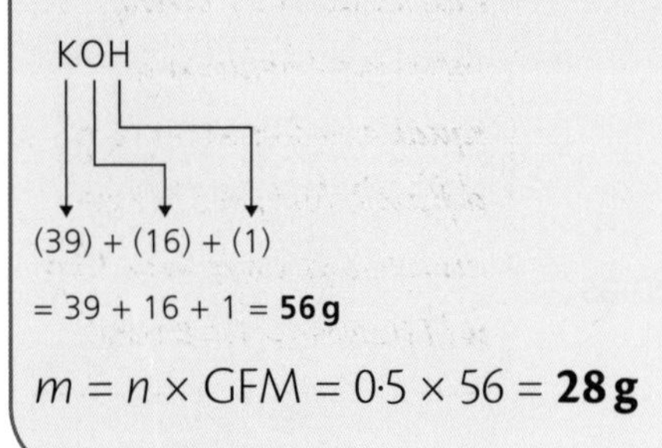

$m = n \times GFM = 0{\cdot}5 \times 56 =$ **28 g**

Example 3

Calculate the concentration of a solution if 33 g of calcium chloride is dissolved in 600 cm^3 of water.

Calculate the number of moles of calcium chloride.

$CaCl_2$

40 + (35·5 × 2)

= 40 + 71 = **111 g**

$$\text{moles} = \frac{\text{mass}}{\text{GFM}} = \frac{33}{111} = \mathbf{0{\cdot}3}\ \text{moles}$$

This can be used to calculate the concentration.

$n = 0{\cdot}3$ moles

$V = 600\ cm^3 = 0{\cdot}6\ l$

$$C = \frac{n}{V} = \frac{0{\cdot}3}{0{\cdot}6} = 0{\cdot}5\ \text{mol}\ l^{-1}$$

Balancing equations

A chemical equation is said to be **balanced** when there are equal amounts of each element on either side of the equation.

The equation shown below is *not* balanced, because there are two oxygens on the left and only one on the right. Atoms cannot magically appear or disappear, so we must balance the equation.

$$Ca + O_2 \rightarrow CaO$$

To balance this equation, we follow these steps:

1. Check that all the formulae are correct.
2. Deal with one element at a time. For example, there is only one calcium on the left and only one on the right, therefore the calciums are balanced.
3. If balancing is required, put the number in front of the substance. For example, there are two oxygens on the left and only one on the right. Therefore, we multiply the compound on the right by two:

   $$Ca + O_2 \rightarrow \mathbf{2}CaO$$

4. Check each element again and repeat step three if required. There are now two calciums on the right and only one on the left. Therefore, we multiply the calcium on the left by two:

   $$\mathbf{2}Ca + O_2 \rightarrow \mathbf{2}CaO$$

The equation is now balanced!

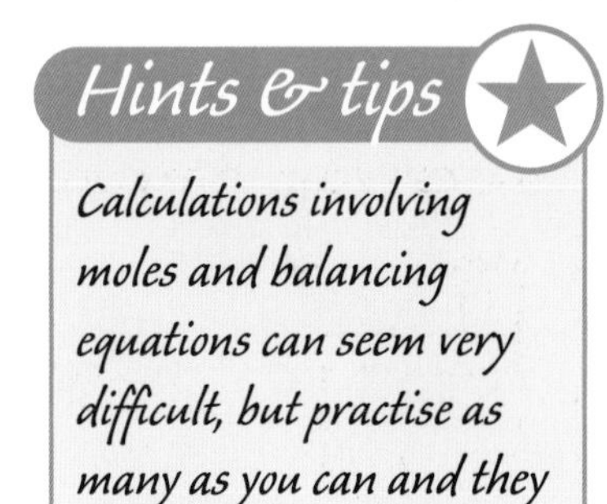

Calculations involving balanced equations

Balanced equations can be used to predict the mass of a product produced by a chemical reaction or the mass of reactants required for the reaction.

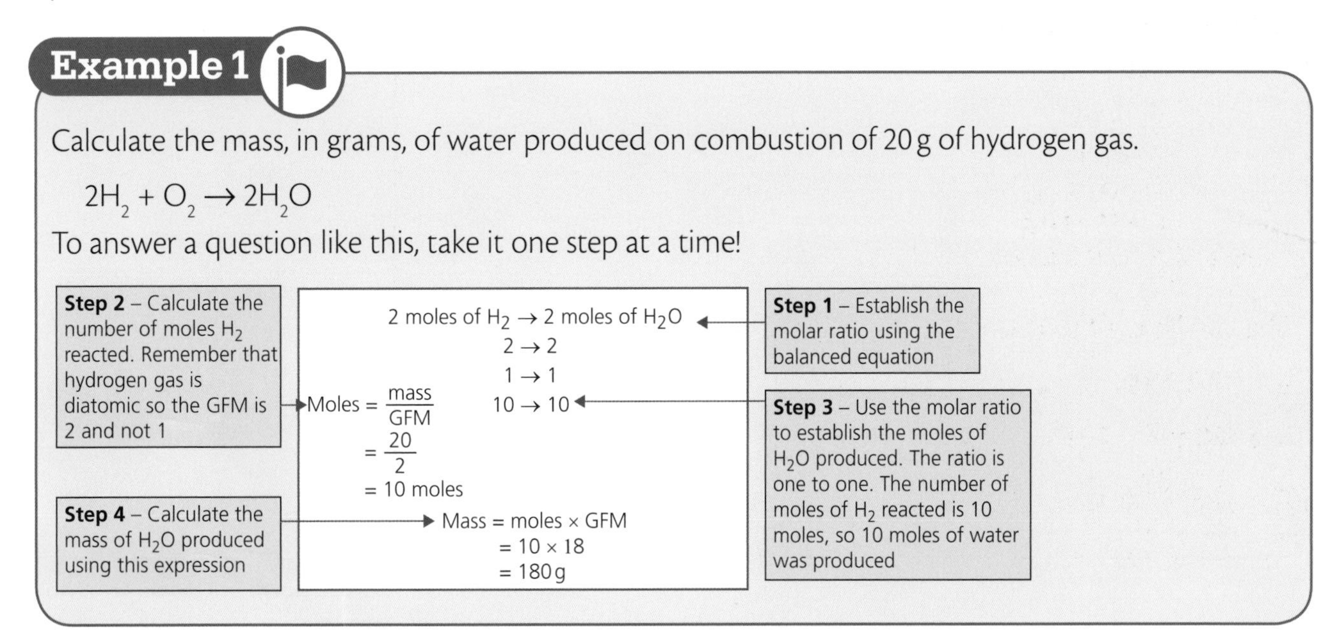

Example 2

Calculate the mass, in grams, of carbon monoxide required to react with 40 g of iron(III) oxide.

$Fe_2O_3 + 3CO \rightarrow 2Fe + 3CO_2$

1 mole of $Fe_2O_3 \rightarrow$ 3 moles of CO

$$\text{moles} = \frac{\text{mass}}{\text{GFM}} = \frac{40}{160} = 0{\cdot}25 \text{ moles}$$

$$1 \rightarrow 3$$

$$0{\cdot}25 \rightarrow 0{\cdot}75$$

$$\text{mass} = \text{moles} \times \text{GFM} = 0{\cdot}75 \times 28 = 21\,\text{g}$$

Hints & tips

If you get questions like these in the exam, you will be provided with the balanced equation.

Chapter 1.7

Percentage composition calculations

The percentage of elements in a compound (percentage composition) can be calculated using this equation:

$$\textbf{percentage compostion} = \frac{\text{mass of element}}{\text{formula mass}} \times 100$$

This equation can be found on page 3 of the Data Booklet.

Example

Calculate the percentage of nitrogen in ammonium nitrate, NH_4NO_3.

$$\text{percentage compostion} = \frac{\text{mass of element}}{\text{formula mass}} \times 100$$

Calculate the formula mass of ammonium nitrate.

NH_4NO_3

$14 + (1 \times 4) + 14 + (16 \times 3)$

$= 14 + 4 + 14 + 48 = \mathbf{80\,g}$

Calculate how much of this is composed of nitrogen.

$2 \times 14 = 28$

Use the formula shown above, substituting in the numbers we have calculated:

$$\text{percentage compostion} = \frac{28}{80} \times 100 = 35\%$$

Study questions

1 a) Write the formula of each of these compounds:
- **i** magnesium oxide
- **ii** calcium chloride
- **iii** aluminium sulfide
- **iv** sodium fluoride
- **v** calcium nitrate
- **vi** ammonium sulfate.

b) Write the formula, showing the charge on each ion, for each of the compounds.

c) Calculate the gram formula mass of each of the compounds.

2 Balance each of the following equations.

a) $Mg + O_2 \rightarrow MgO$

b) $CH_4 + O_2 \rightarrow CO_2 + H_2O$

c) $Na + O_2 \rightarrow Na_2O$

⇨

d) $Li + H_2O \rightarrow LiOH + H_2$

e) $Fe + O_2 \rightarrow Fe_2O_3$

f) $N_2 + H_2 \rightarrow NH_3$

3 A student has been asked to prepare 250 cm^3 of 2 mol l^{-1} sodium hydroxide solution. Calculate the mass, in grams, of sodium hydroxide, NaOH, required to produce this solution.

4 Potassium nitrate can be used as fertiliser. To test its effectiveness a farmer decides to use a 1 mol l^{-1} solution on his test crops. Calculate the mass, in grams, of potassium nitrate, KNO_3, required to produce 1·5 litres of 1 mol l^{-1} solution.

5 Calculate the concentration, in mol l^{-1}, of a magnesium bromide solution, $MgBr_2$ (aq), if 20·9 g is dissolved in 50 cm^3 of water.

6 The synthetic fertiliser ammonium phosphate $(NH_4)_3PO_4$, is commonly used by gardeners as a fertiliser for tomato plants.

a) Calculate the percentage of nitrogen in ammonium phosphate.

b) Calculate the percentage of phosphorus in ammonium phosphate.

7 Calculate the mass, in grams, of carbon dioxide produced on combustion of 32 g of methane gas.

$$CH_4 + 2O_2 \rightarrow CO_2 + 2H_2O$$

8 Iron(II) sulfate can be used in the treatment of anaemia. Iron(II) sulfate can be produced by reacting iron(II) oxide with sulfuric acid.

If 25 g of iron(II) oxide is reacted with excess sulfuric acid, calculate the mass of iron(II) sulfate that will be produced.

$$FeO + H_2SO_4 \rightarrow FeSO_4 + H_2O$$

Chapter 1.8
Acids and bases

The pH scale gives us an indication of how acidic or alkaline a substance is. In this chapter we will revise what makes a substance acidic or alkaline, and what happens when we add acids and alkalis together.

The pH scale

Before we progress through this chapter it is important to have a look at the **pH** scale.

The pH scale is a measure of the concentration of hydrogen ions (**p**otential **H**ydrogen) and it is a continuous scale from below 0 to above 14. (We will look at hydrogen ions later in this chapter.)

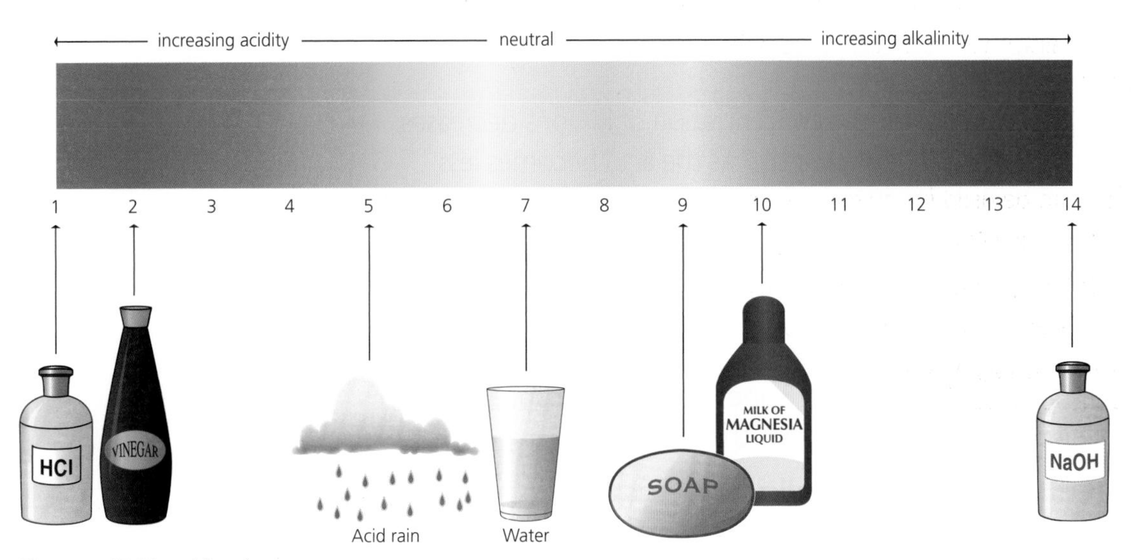

Figure 1.49 The pH scale

Key points

- A pH of *below* 7 indicates an acidic solution. (pH paper or universal indicator solution, turns red or orange)
- A pH of *above* 7 indicates an alkaline solution. (pH paper or universal indicator solution, turns blue or purple)
- A pH of 7 is a neutral solution, such as pure water. (pH paper or universal indicator solution, turns green)

Acidic solutions

Acidic solutions have a pH of between 1 and 6. This is because they contain hydrogen (H^+) ions in a greater concentration than hydroxide (OH^-) ions.

Hydroxide (OH^-) ions are the opposite of H^+ ions and cancel them out (see neutralisation, page 39). So, for a solution to be acidic it must have a higher concentration of H^+ ions than OH^- ions.

The presence of H^+ ions explains why **acids** react as they do, and all acidic solutions contain H^+. All acids contain ions and as a result they conduct electricity, because the ions are free to carry the charge.

Remember

The table below lists the commonly used lab acids and ethanoic acid (see page 60). It is important to learn the names and formulae of these acids.

Acid name	Formula	Ionic formula
hydrochloric acid	HCl	$H^+(aq)\ Cl^-(aq)$
sulfuric acid	H_2SO_4	$(H^+)_2(aq)\ SO_4^{2-}(aq)$
nitric acid	HNO_3	$H^+(aq)\ NO_3^-(aq)$
ethanoic acid	CH_3COOH	$H^+(aq)\ CH_3COO^-(aq)$

Note that the H^+ ions are present in the ionic formula of all acids.

When an acid is diluted with water, the concentration of H^+ ions decreases; this increases the pH of the solution towards 7 as the acid becomes less acidic. This dilution would also reduce the conductivity of the acid.

Alkali solutions

Alkaline solutions have a pH of between 8 and 14. This is because they contain a greater concentration of OH^- ions than H^+ ions.

The presence of OH^- ions explains why **alkalis** react as they do, and all alkaline solutions contain OH^-. Like acids, alkalis conduct electricity because they have ions that are free to carry the charge.

Remember

The table below lists the commonly used lab alkalis. It is important to learn the names and formulae of these alkalis.

Alkali name	Formula	Ionic formula
sodium hydroxide	$NaOH$	$Na^+(aq)\ OH^-(aq)$
calcium hydroxide	$Ca(OH)_2$	$Ca^{2+}(aq)\ (OH^-)_2(aq)$
lithium hydroxide	$LiOH$	$Li^+(aq)\ OH^-(aq)$

Note that the OH^- ions are present in the ionic formula of all alkalis.

When an alkali is diluted with water, the concentration of OH^- ions decreases; this decreases the pH of the solution towards 7 as the alkali becomes less alkaline. It also reduces the conductivity of the alkali.

Making acids and alkalis

Dissolving different soluble oxides in water forms acidic and alkaline solutions. The solubility of oxides can be found on page 8 of the Data Booklet.

Key points 

- When a soluble non-metal oxide is dissolved in water an acidic solution is produced.
- When a soluble metal oxide, known as a **base**, is dissolved in water an alkaline solution is produced.

Metal oxides such as Li_2O, Na_2O and K_2O, dissolve in water to produce alkaline solutions (solutions with a pH greater than 7).

$$Li_2O + H_2O \rightarrow 2LiOH$$

Looking at page 8 of the Data Booklet, you will notice that only the oxides of group 1 (and some of group 2) are soluble; most other metal oxides are insoluble. So, for example, adding zinc oxide to water does not produce zinc hydroxide because the zinc oxide would not dissolve in the water.

Non-metal oxides such as sulfur dioxide and carbon dioxide dissolve in water to form acidic solutions (solutions with a pH below 7).

$$CO_2 + H_2O \rightleftharpoons H_2CO_3$$

Neutral solutions

A neutral solution has an equal concentration of hydrogen and hydroxide ions, which cancel each other out. This can be shown as an equation:

$$H^+ + OH^- \rightleftharpoons H_2O$$

You will notice that the reaction is reversible. This is because a very small proportion of water molecules naturally **dissociate** (break up) to form an equal number of hydrogen and hydroxide ions. The dissociation of water is the reason that water can conduct electricity, although poorly; there are a small number of hydrogen and hydroxide ions that are free to carry the charge.

Key points 

- Neutral solutions contain equal amounts of hydrogen and hydroxide ions.
- A small amount of water molecules dissociate to form hydrogen and hydroxide ions.

Neutralisation

Neutralisation is the reaction of an acid with a base, such as an alkali, metal oxide, metal hydroxide or carbonate. This moves the pH of the acid upwards towards 7 and the pH of the alkali downwards towards 7.

Neutralisation reactions occur every day, for example, in the treatment of acid indigestion or when adding lime to lakes to reduce their acidity and counteract the effects of acid rain.

Acids are neutralised by alkalis and bases to form **salt** and water.

acid + alkali → salt + water

acid + base → salt + water

Metal oxides, metal hydroxides, metal carbonates and ammonia neutralise acids and are called bases. Those bases that are soluble dissolve to form alkaline solutions.

It's important to note that if the base is a metal carbonate then carbon dioxide gas would also be produced.

acid + metal carbonate → salt + water + carbon dioxide

The name of the salt produced depends on the acid and alkali that reacted. Naming the salt is straightforward if you learn the information in the table below.

Remember

It is important to learn the names of the salts produced when the acids in the table are neutralised.

Acid name	Salt name ends in
hydrochloric acid	chloride
sulfuric acid	sulfate
nitric acid	nitrate

Example

1 hydrochloric acid + sodium hydroxide → sodium chloride + water

2 nitric acid + calcium carbonate → calcium nitrate + water + carbon dioxide

3 sulfuric acid + magnesium oxide → magnesium sulfate + water

The salt is produced when the hydrogen ion of the acid is replaced by the metal ion or the ammonium ion of the base, but why is water produced?

During neutralisation, the H^+ ion from the acid joins with the OH^- ion from the alkali, explaining why water is formed in these reactions:

$$H^+ + OH^- \rightarrow H_2O$$

This can be illustrated clearly using ionic equations.

Ionic equations

Previously you learned the ionic formulae of acids and alkalis. These can now be used to form ionic equations.

Example 1

Word equation:

hydrochloric acid + sodium hydroxide ⟶ sodium chloride + water

Formula equation:

$HCl + NaOH \longrightarrow NaCl + H_2O$

The ionic equation simply uses the ionic formulae of the substances shown in the formula equation.

Ionic equation:

$H^+(aq)\ Cl^-(aq) + Na^+(aq)\ OH^-(aq) \longrightarrow Na^+(aq)\ Cl^-(aq) + H_2O(l)$

This can be shortened further by removing the **spectator ions**. The spectator ions are ions that are present during the reaction, but are unchanged by the reaction.

$H^+(aq)\ Cl^-(aq) + Na^+(aq)\ OH^-(aq) \longrightarrow Na^+(aq)\ Cl^-(aq) + H_2O(l)$

When the spectator ions are removed we are left with a very familiar equation:

$H^+ + OH^- \longrightarrow H_2O$

Remember

A spectator ion is like a spectator at a football match. They are there at the match but are not taking part in the game. They are simply spectators watching the reaction take place.

Acids can also react with other bases, such as metal carbonates.

When an acid reacts with any carbonate, carbon dioxide is produced.

Example 2

$$2HCl + CaCO_3 \rightarrow CaCl_2 + H_2O + CO_2$$

The same method as before is used to name the salt and we can also write an ionic equation for this reaction:

$$2H^+ (aq)\ 2Cl^- (aq) + Ca^{2+}(aq)\ CO_3^{2-}(aq) \rightarrow Ca^{2+}(aq)\ 2Cl^- (aq) + H_2O(l) + CO_2(g)$$

(Note that CO_2 and H_2O are covalent so they can't have an ionic formula!)

When the spectator ions are removed from this equation we are left with:

$$2H^+ (aq) + CO_3^{2-}(aq) \rightarrow H_2O(l) + CO_2(g)$$

It is important to note here that for an ion to be classed as a spectator ion it must be completely unchanged by the reaction.

Example 3

$$NaI\ (aq) + AgNO_3\ (aq) \rightarrow NaNO_3(aq) + AgI(s)$$

In this example Ag^+ and I^- are *not* spectator ions as they have changed state from being in solution to being a solid.

Titrations

An acid–base **titration** procedure is used to determine the concentration of an acid. A pipette is used to accurately measure the volume of a standard solution of an acid, such as sulfuric acid.

A **standard solution** is a solution of accurately known concentration.

This is then reacted with a base, which is added to the acid via a burette until the acid is neutralised. An indicator is usually added to provide a colour change at the point when neutralisation is complete. This is called the end-point.

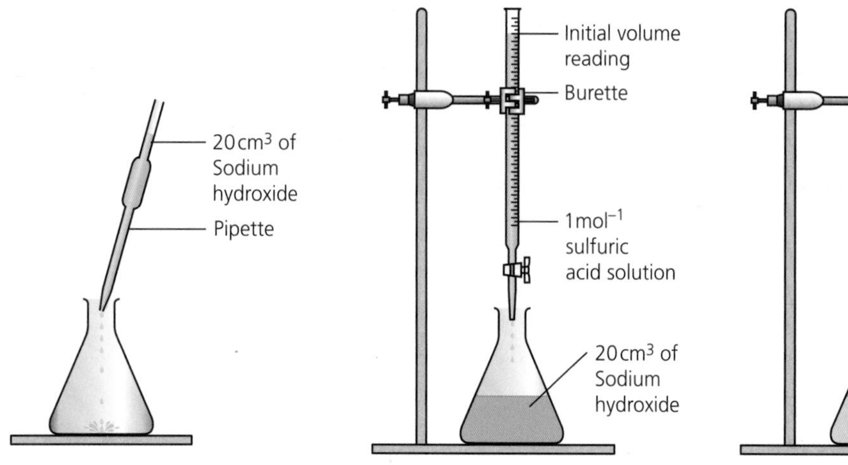

Figure 1.50 Equipment for a titration experiment

This process is repeated until concordant results are obtained. The term 'concordant' means results that agree with each other, and for National 5 concordant results are within 0·2 cm^3 of each other.

Example 1

Titration	Start volume (cm^3)	Final volume (cm^3)	Total volume (cm^3)
1	0·0	10·4	10·4
2	11·0	20·9	9·9
3	21·0	31·1	10·1

Only concordant results should be used to calculate the average volume of acid added.

The results of titrations 2 and 3 are concordant (within 0·2 cm^3 of each other) and only these results should be used to calculate the average volume of acid added.

$$\text{average volume} = \frac{9{\cdot}9 + 10{\cdot}1}{2} = 10{\cdot}0\ cm^3$$

This average volume can be used to calculate the concentration of the alkali that was in the conical flask, as shown in Example 2.

Example 2

The equation for the reaction in Figure 1.50 is:

$H_2SO_4 + 2NaOH \rightarrow Na_2SO_4 + 2H_2O$

Calculate the concentration, in $mol\ l^{-1}$, of the sodium hydroxide solution.

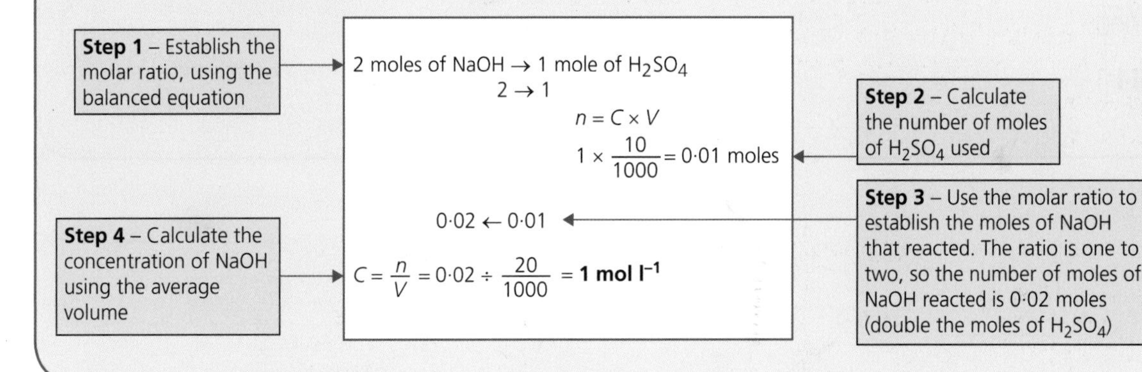

Example 3 

A student carried out titrations to establish the concentration of a sodium hydroxide solution and recorded their results in the table.

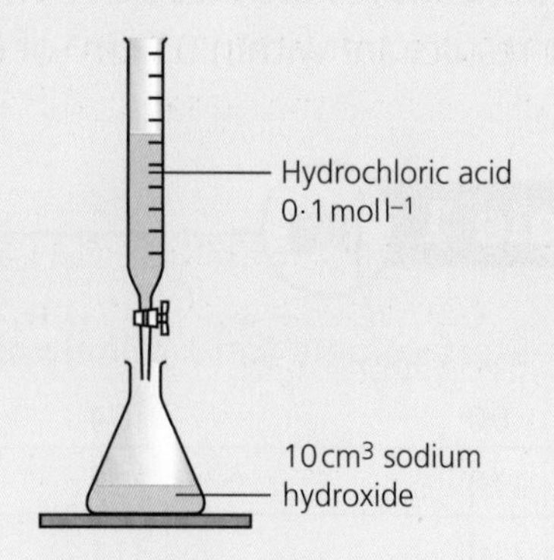

Figure 1.51 The student used this equipment for their titration experiment

Titration	Initial burette reading (cm^3)	Final burette reading (cm^3)	Total (cm^3)
1	0·0	15·7	15·7
2	16·0	30·9	14·9
3	31·0	46·0	15·0

Calculate the average volume of acid, in cm^3, that should be used in calculating the concentration of the sodium hydroxide solution.

$$\text{average volume} = \frac{14{\cdot}9 + 15{\cdot}0}{2} = 14{\cdot}95\,cm^3$$

The equation for the reaction is:

$$HCl(aq) + NaOH(aq) \rightarrow NaCl(aq) + H_2O(l)$$

Calculate the concentration, in $mol\,l^{-1}$, of the sodium hydroxide solution.

1 mole of HCl $\rightarrow$ 1 mole of NaOH

$1 \rightarrow 1$

$n = C \times V$

$$0{\cdot}1 \times \frac{14{\cdot}95}{1000} = 0{\cdot}0015 \text{ moles}$$

0·0015 mole of HCl $\rightarrow$ 0·0015 mole of NaOH

$$C = \frac{n}{V} = \frac{0{\cdot}0015}{\frac{10{\cdot}0}{1000}}$$

$= \mathbf{0{\cdot}15\ mol\ l^{-1}}$

Titrations can be used to produce soluble salts. When the titration is complete and concordant results are obtained, the process can be repeated without an indicator to produce an uncontaminated salt solution. A dry sample of the salt can be obtained by evaporating off the water.

Insoluble metal carbonates and metal oxides can be used to produce soluble salts. Copper sulfate can be produced by reacting copper carbonate with sulfuric acid.

$$CuCO_3 + H_2SO_4 \rightarrow CuSO_4 + H_2O + CO_2$$

To obtain a dry sample of copper sulfate crystals, the following three steps must be performed.

1 Copper carbonate is added to a measured volume of sulfuric acid until no more carbon dioxide is given off.
2 The excess copper carbonate is removed by **filtration**. Filtration is a method used to separate an insoluble solid from a liquid (Figure 1.52). The **residue** is left on the filter paper and the **filtrate** is the liquid that is collected after filtration.
3 A dry sample of copper sulfate is obtained by evaporating off the water. Evaporation is a method used to separate a soluble solid from a liquid (Figure 1.53).

This method can also be used to produce soluble salts from metals.

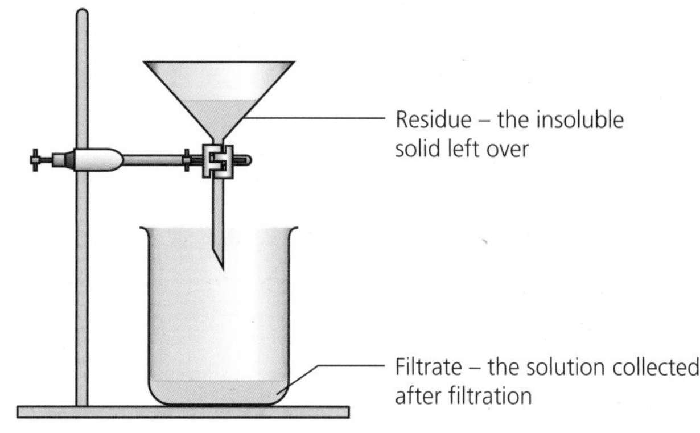

Figure 1.52 Filtration

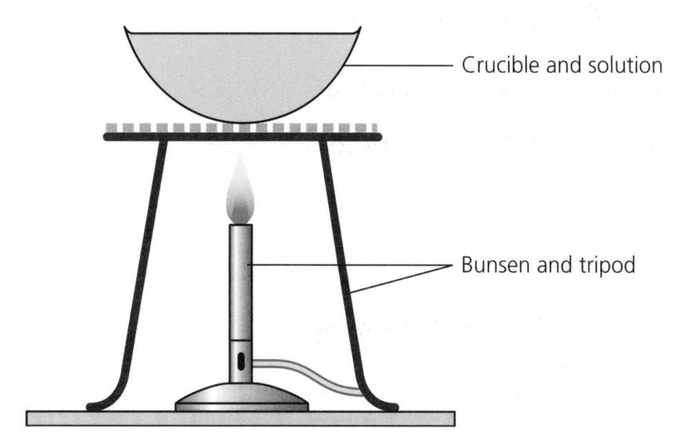

Figure 1.53 Evaporation

Precipitation

Precipitation is the formation of an insoluble solid, a **precipitate**, by reacting two liquids together. A precipitate is always insoluble. Page 8 of your Data Booklet lists the solubility of selected compounds.

lead nitrate + sodium iodide → sodium nitrate + lead iodide

The precipitate is lead iodide, an insoluble solid.

The precipitate can easily be identified from formula equations which include state symbols.

$$Pb(NO_3)_2(aq) + 2NaI(aq) \rightarrow 2NaNO_3(aq) + PbI_2(s)$$

Lead iodide is the precipitate and it is shown in the solid state.

Study questions

1 0·75 mol of citric acid was dissolved in 250 cm³ of water. Which of the following is the concentration of the solution formed?
a) $0{\cdot}003\ mol\ l^{-1}$
b) $0{\cdot}3\ mol\ l^{-1}$
c) $3\ mol\ l^{-1}$
d) $333\ mol\ l^{-1}$

2 Which of the following compounds is classed as a salt?
a) sodium choride
b) calcium oxide
c) sulfur dioxide
d) hydrogen chloride

3 Which of the following oxides dissolves in water to produce an acidic solution?
a) SiO_2
b) SO_2
c) Fe_2O_3
d) PbO_2

4 An unknown white solid is soluble in water. The unknown solid also reacts with hydrochloric acid to produce carbon dioxide gas. The solid could be
a) sodium oxide
b) copper(II) carbonate
c) aluminium oxide
d) potassium carbonate.

5 Identify the spectator ions in the following reaction.

$$H^+ (aq)\ NO_3^- (aq) + K^+ (aq)\ OH^- (aq) \rightarrow K^+ (aq)\ NO_3^- (aq) + H_2O(l)$$

a) H^+ and NO_3^-
b) H^+ and OH^-
c) K^+ and NO_3^-
d) K^+ and OH^-

6 Which of the following oxides, when shaken with water, would leave the pH unchanged? You may wish to use your Data Booklet to help you.
a) carbon dioxide
b) copper oxide
c) sodium oxide
d) sulfur dioxide

7 A technician dissolves solid sodium oxide in water to produce a concentrated solution of sodium hydroxide.
a) Write the word equation for this reaction.
b) Write the equation using symbols and formulae.
c) Suggest a pH for the solution produced.
d) Suggest why using this method to produce a solution of iron hydroxide would be difficult.

8 The concentration of ethanoic acid in vinegar can be determined by titrating a 10 cm³ sample of the acid with 0·5 mol l⁻¹ sodium hydroxide solution.

The equation for the reaction is:

$$CH_3COOH(aq) + NaOH(aq) \rightarrow CH_3COONa(aq) + H_2O(aq)$$

The results for the titration were recorded in the table below.

Titration	Volume of sodium hydroxide (cm³)
1	21·3
2	20·1
3	19·9

a) Calculate the concentration, in $mol\ l^{-1}$, of the ethanoic acid in the vinegar.

b) Write the ionic formula for ethanoic acid.

c) Name a spectator ion in the reaction.

9 Ammonium sulfate was prepared by reacting ammonia with sulfuric acid.

$$H_2SO_4(aq) + 2NH_3(aq) \rightarrow (NH_4)_2SO_4(aq)$$

0·1 $mol\ l^{-1}$ sulfuric acid was added to 10 cm^3 of 0·1 $mol\ l^{-1}$ ammonia solution.

Calculate the volume, in cm^3, of acid required to neutralise the ammonia solution.

Section 2 Nature's chemistry

Chapter 2.1
Homologous series

Hydrocarbons are compounds made from carbon and hydrogen only. In this chapter we will discover that there are families of hydrocarbons. These families are known as homologous series.

Key points 

- A **hydrocarbon** is a compound made up of carbon and hydrogen only.
- A homologous series is a family of hydrocarbons with similar chemical properties, and that can be represented by a general formula.

To explain what is meant by the term **homologous series** we can look at a human family.

John McBride is the oldest member of the McBride family and he has one sister called Grace McBride and one brother called Joseph McBride.

The children all belong to the same family, which is why they share the same name ending: McBride. Although they are all in the same family, so are similar in some ways, they look a bit different from one another and act in slightly different ways.

The McBrides resemble any homologous series. The compounds in a homologous series all have the same ending to their names and each member has similar chemical properties.

The simplest homologous series is called the **alkanes**.

Figure 2.1 The McBride family – a homologous series!

Alkanes

All alkanes all have the same name ending: –**ane**, for example methane.

The alkanes are insoluble in water but have very important uses, such as those shown in Figure 2.2.

Figure 2.2 Alkanes include methane which is used for cooking and heating, and octane which is used as petrol for cars

Remember

Here are the names, molecular and structural formulae of the first four members of the alkane homologous series. You must learn the names, molecular and structural formulae of the first **eight** alkanes. The names are listed on page 9 of the Data Booklet.

Name	Molecular formula	Full structural formula	Shortened structural formula
methane	CH_4	H | H—C—H | H	CH_4
ethane	C_2H_6	H H | | H—C—C—H | | H H	CH_3CH_3
propane	C_3H_8	H H H | | | H—C—C—C—H | | | H H H	$CH_3CH_2CH_3$
butane	C_4H_{10}	H H H H | | | | H—C—C—C—C—H | | | | H H H H	$CH_3CH_2CH_2CH_3$

Hints & tips

To make it easy to remember all the alkanes (as well as other homologous series), use this little mnemonic. Try it, or make up your own to remember the names.

Monkeys	*Methane*	CH_4
Eat	*Ethane*	C_2H_6
Peanut	*Propane*	C_3H_8
Butter	*Butane*	C_4H_{10}
Perhaps	*Pentane*	C_5H_{12}
Harry	*Hexane*	C_6H_{14}
Heptane	*Heptane*	C_7H_{16}
Objects	*Octane*	C_8H_{18}

Figure 2.3 Monkeys eat peanut butter, perhaps Harry Heptane objects!

All homologous series have a general formula. The general formula allows you to work out the molecular formula of any member of the series.

The general formula of the alkanes is: $\mathbf{C_nH_{2n+2}}$

where n is the number of carbon atoms.

Example

Give the molecular formula of pentane.

Using your mnemonic, you can work out that pentane is the fifth alkane and, therefore, has five carbon atoms ($n = 5$). To calculate the number of hydrogen atoms use the formula:

$$C_nH_{2n+2} = C_5H_{(2 \times 5)+2}$$

So the molecular formula of pentane is $\mathbf{C_5H_{12}}$.

As with all homologous series, there is a regular change in chemical and physical properties with increasing numbers of carbon atoms. For example, the boiling point of the alkanes increases, but the flammability decreases as the carbon chain increases.

Alkane	Boiling point (°C)
methane	−162
ethane	−89
propane	−42
butane	−1
pentane	36

Key point

* As the length of the carbon chain increases more intermolecular bonds (bonds between molecules) form, resulting in higher boiling points and melting points, but lower flammability.

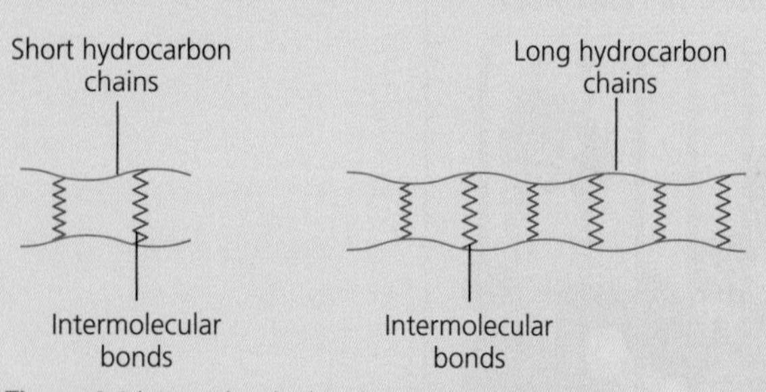

Figure 2.4 Intermolecular bonds between carbon chains of different lengths

Alkenes

The second homologous series is the **alkenes**. All alkenes have the ending **–ene**, for example ethene. Alkenes are used to make polymers (see page 79) and alcohols, and are insoluble in water.

They are different from the alkanes because they contain a carbon-to-carbon double bond.

Remember

Here are the names, molecular and structural formulae of the first three members of the alkene homologous series.

Name	Molecular formula	Full structural formula
ethene	C_2H_4	H H C=C H H
propene	C_3H_6	H H H H—C—C=C H H
*but-1-ene	C_4H_8	H H H H H—C—C—C=C H H H

***Page 55 describes why but-1-ene is not simply called butene.**

Key point 

- There is no methene because there must be at least two carbon atoms to form a carbon-to-carbon double bond.

The alkenes have the general formula: $\mathbf{C_nH_{2n}}$
where n is the number of carbon atoms.

Cycloalkanes

The third homologous series is the **cycloalkanes**. The cycloalkanes are used as solvents and **fuels** and, like the alkanes and alkenes, are insoluble in water.

All cycloalkane names start with **cyclo**- and have the ending –**ane**, for example cyclopropane.

Remember

The names, molecular and structural formulae of the first three cycloalkanes are shown in the table.

Name	Molecular formula	Full structural formula
cyclopropane	C_3H_6	H H C H—C—C—H H H
cyclobutane	C_4H_8	H H H—C—C—H H—C—C—H H H
cyclopentane	C_5H_{10}	H H H C—C H H—C C—H H C H H H

The cycloalkanes have the same general formula as the alkenes: $\mathbf{C_nH_{2n}}$ where n is the number of carbon atoms.

Saturated or unsaturated?

The alkenes are described as being **unsaturated hydrocarbons**. This means that they contain a carbon-to-carbon double bond.

This double bond makes alkenes more reactive than alkanes, which have only single bonds. The alkanes and cycloalkanes are described as saturated because they only contain single carbon-to-carbon bonds.

- Compounds that contain a carbon-to-carbon double (or triple) bond are described as being unsaturated.
- **Saturated hydrocarbons** contain single carbon-to-carbon bonds only.

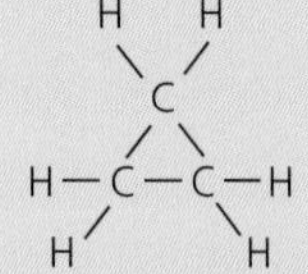

Propane – saturated   Cyclopropane – saturated   Propene – unsaturated

Figure 2.5 Saturated and unsaturated hydrocarbons

Addition reactions

Alkenes, because they are unsaturated, are more reactive than alkanes and cycloalkanes.

The double carbon-to-carbon bond allows them to take part in **addition reactions** – which involve the addition of a molecule, such as water, or a diatomic molecule, such as hydrogen, across the double bond.

Example

This is the reaction of ethene with hydrogen.

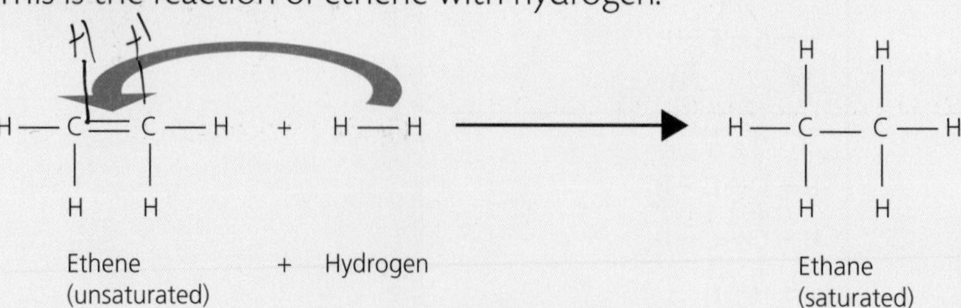

Figure 2.6 The reaction of ethene with hydrogen

The hydrogen molecule attacks the double bond of the ethene molecule, and adds across it to produce ethane. Adding hydrogen across a double bond changes the unsaturated alkene into the saturated alkane.

alkene + hydrogen $\rightarrow$ alkane

The reaction between an alkene and hydrogen is an addition reaction, known as **hydrogenation**.

Alkenes can be identified because they have this capability for addition reactions. When bromine, which is orange in colour, is added to an alkene it is immediately decolourised. This doesn't happen with a saturated hydrocarbon like an alkane or cycloalkane, because these don't have a double bond for the bromine to add across. If a carbon-to-carbon double bond is present the bromine will be decolourised.

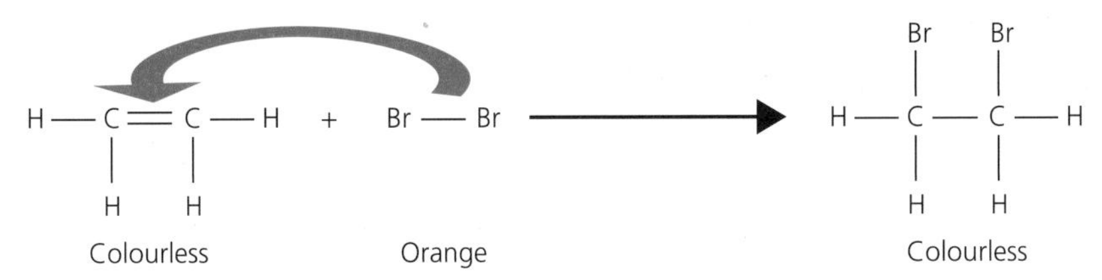

Figure 2.7 The reaction of ethene with bromine

The reaction between an alkene and a **halogen** (a group 7 element) forms **dihaloalkanes** such as the colourless molecule shown in the equation above.

Addition reactions can also be used to form **alcohols**.

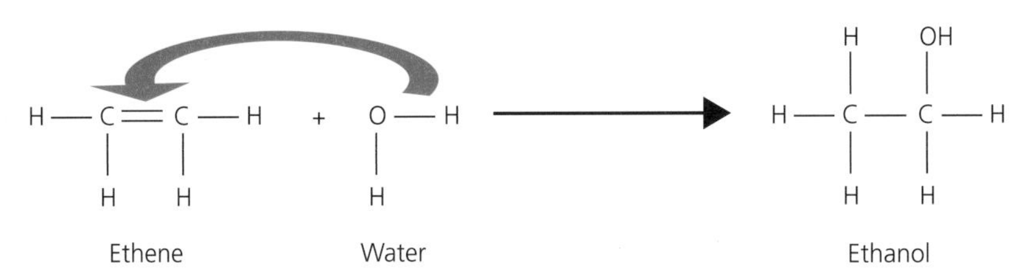

Figure 2.8 The reaction between ethene and water to form ethanol

An 'H' from the water adds to one of ethene's carbons and the remaining 'OH' adds to the other, forming an alcohol. This reaction is known as **hydration**.

Branched hydrocarbons

Figure 2.9 shows a branched chain hydrocarbon with the name 3-methylpentane. Branched-chain hydrocarbons are simply hydrocarbons that have a branch or branches coming off the main chain.

Branched hydrocarbon names seem to be very complicated, but the naming rules are, in fact, simple.

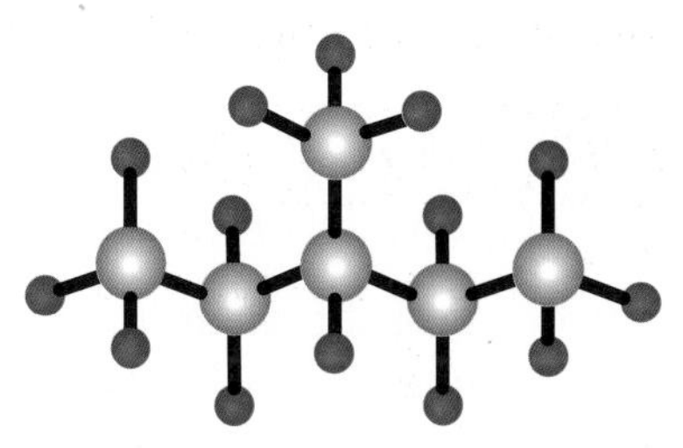

Figure 2.9 3-methylpentane

Key points

1. Find the longest carbon chain and name as normal, for example five carbon atoms in the longest chain means that it is a type of pentane.
2. Next, identify the branch and name that, for example methyl (one carbon) or ethyl (two carbons). In Figure 2.10, R stands for the long chain, which can be any length.

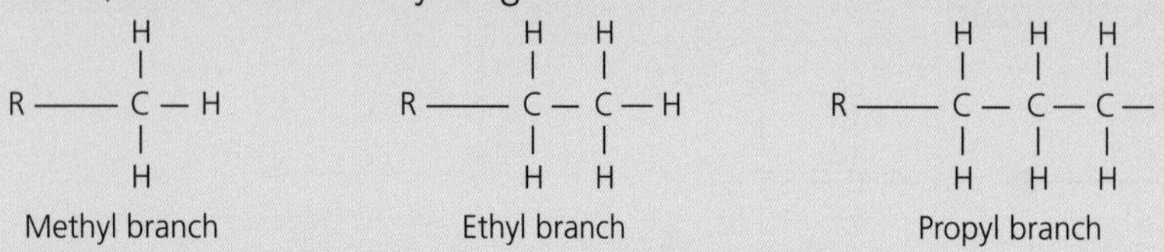

Figure 2.10 Three examples of branches of hydrocarbons

3. Number the carbons in the long chain so that the branch is on the lowest possible number.

Example 1

State the name of this branched hydrocarbon.

```
    H   H   H   H   H              H   H   H   H   H
    |   |   |   |   |              |   |   |   |   |
H — C — C — C — C — C — H      H — C — C — C — C — C — H
    |   |   |   |   |              |   |   |   |   |
    H   H   H   |   H              H   H   H   |   H
                |                           ┌──|──────┐
            H — C — H                       │H—C—H    │
                |                           │  |      │
                H                           │  H      │
                                            └─────────┘
```

1. The longest chain has five carbons (shown by the red line).
2. The branch has one carbon (shown by the green box).
3. The branch is on carbon 2 (not carbon 4, because you always use the lowest number).

So the name is **2-methylpentane**

The shortened structural formula of 2-methylpentane is $CH_3CH_2CH_2CH(CH_3)CH_3$.

The brackets shown around the CH_3 indicate that this is a branch, not part of the main chain. You should know how to use brackets correctly in structural formulae, for the exam.

There can be more than one branch. If more than one branch is present, then a prefix must be used:

- di = two identical branches
- tri = three identical branches
- tetra = four identical branches

Example 2 

Name the following branched hydrocarbon.

```
    H    H    H    H    H    H    H
    |    |    |    |    |    |    |
H — C —  C —  C —  C —  C —  C —  C — H
    |    |    |    |    |    |    |
    H    |    H    H    |    H    H
         |              |
     H — C — H      H — C — H
         |              |
         H              H
```

Shortened structural formula:
$CH_3CH(CH_3)CH_2CH_2CH(CH_3)CH_2CH_3$

A molecule like this may seem a bit scary, but stick to the rules and it will be easy to name!

1. The longest chain has 7 carbons (shown by the red line), so this is a type of heptane.
2. There are two methyl branches (shown by the green boxes).
3. The branches are on carbons 2 and 5.

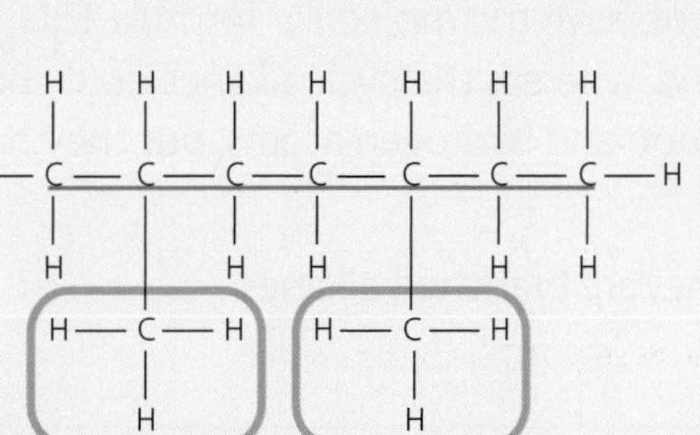

So the name is **2,5-dimethylheptane**.

The same rules apply when naming alkenes – with one difference: the position of the double bond must be identified. The number showing the position, which is the lowest number possible, is placed before the '–ene' ending to the name.

Example 3 

State the name of this hydrocarbon.

```
    H    H    H    H
    |    |    |    |   / H
H — C —  C —  C —  C = C
    |    |    |        \ H
    H    H    |
          H — C — H
              |
              H
```

Shortened structural formula: $CH_3CH_2CH(CH_3)CHCH_2$

3-methylpent-1-ene

Isomers

Example 1

Cyclopropane is an isomer of propene.

ISOMERS

Propene

Cyclopropane

Figure 2.11 Isomers propene and cyclopropane

Both cyclopropane and propene have the molecular formula C_3H_6, but the alkenes have a double bond, whereas the cycloalkanes don't. Both have the same number of carbon and hydrogen atoms, but they have different structures.

Isomers of straight chain alkanes are branched alkanes.

Example 2

2-methylpentane is an isomer of hexane.

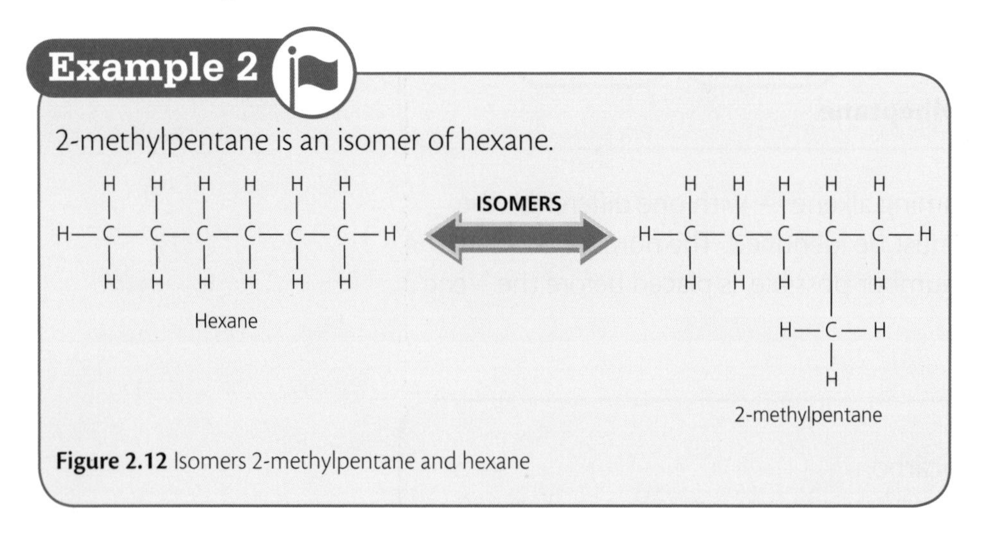

Figure 2.12 Isomers 2-methylpentane and hexane

Isomers have different physical properties because they have different structures. This difference in physical properties for hexane and 2-methylpentane is illustrated in the table.

Property	Hexane	2-methylpentane
melting point	−95°C	−154°C
boiling point	69°C	60°C
molecular mass	86	86

The different properties are a direct result of the structure of the isomers.

Key points

* **Isomers** are compounds that have the same molecular formula, but a different structural formula.
* This means that they have the same number of carbon and hydrogen atoms but they are arranged differently.

Study questions 

1 Figure 2.13 is the structure of neopentane, which is an extremely flammable gas.

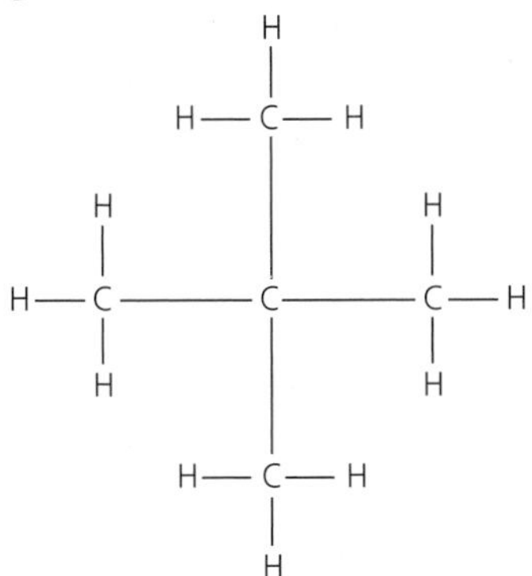

Figure 2.13

Which of the following is the correct systematic name of neopentane?

a) 3,2-dimethylbutane
b) 2,2-dimethylpropane
c) 2,3-dimethylbutane
d) 3,2-dimethylpropane.

2 The first three members of the alkynes homologous series are shown in Figure 2.14.

Ethyne	H—C≡C—H
Propyne	H—C≡C—CH₃ (H—C≡C—C(H)(H)—H)
Butyne	H—C(H)(H)—C≡C—C(H)(H)—H

Figure 2.14

Which of the following is the correct general formula for the alkynes?

a) C_nH_{2n}
b) C_nH_{2n-2}
c) C_nH_{2n+2}
d) C_nH_n

3 Which of the following molecules is an isomer of butane?

a) butene
b) 2-methylpropane
c) cyclobutane
d) 2-methylbutane

4 Which of the following is the correct systematic name for the compound with this shortened structural formula: $CH_3CH(CH_3)CH_2CH_2CH_2CH_3$?

a) heptane
b) hexane
c) 2-methylhexane
d) 2-methylheptane

5 Name the type of reaction represented by the following equation:

$$C_2H_4 + H_2O \rightarrow C_2H_5OH$$

a) addition
b) hydrolysis
c) combustion
d) fermentation

6 Volatile organic compounds (VOCs) are used in paints as solvents and the VOC content is displayed on most paint cans. An example of a VOC compound used in paints is methanal, which is the first member of the aldehydes homologous series.

The structural formula of methanal is shown in Figure 2.15.

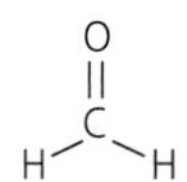

Figure 2.15

a) State what is meant by the term *homologous series.*
b) Write the molecular formula of methanal.

7 A student was testing for unsaturation in four unknown organic compounds. He recorded his results in the table below.

Hydrocarbon	Molecular formula	Observations on adding bromine solution	Saturated or unsaturated?
A	C_5H_{10}		saturated
B	C_5H_{10}		unsaturated
C	C_5H_{12}	remains orange	saturated
D	C_5H_8	orange to colourless	unsaturated

a) Copy and complete the table with the observations that you would expect to see when adding bromine solution to samples of molecules A and B.
b) Draw the full structural formula of molecule B.
c) Molecules A and B have the same molecular formula but different structures. State the name given to two such molecules.

Chapter 2.2
Consumer products

In this chapter we will continue to look the various homologous series. We will specifically look at their importance in the manufacturing of consumer products.

Alcohols

When we think of alcohols, we think of alcoholic drinks, such as beers and wines, but alcohols are actually another homologous series. They have the general formula: $\mathbf{C_nH_{2n+1}OH}$

where *n* is the number of carbon atoms.

The second member of the alcohol homologous series is ethanol.

Molecular formula	C_2H_5OH
Full structural formula	H–C(H)(H)–C(H)(H)–OH
Shortened structural formula	CH_3CH_2OH

Key point

* Alcohols are a family of compounds with a hydroxyl functional group (–OH).

Ethanol is the alcohol that is found in alcoholic drinks. The **functional group** (–OH) is the part of the molecule that dictates the properties of alcohols. In other words, the functional group makes an alcohol function in a typical way. Think of the functional group as being the 'brain' of the molecule, controlling the way it reacts.

As with all homologous series, there are many members and we use the same rules as before to name them. The position of the functional group does not have to be identified when naming methanol and ethanol as it can only be on the first carbon atom.

Example

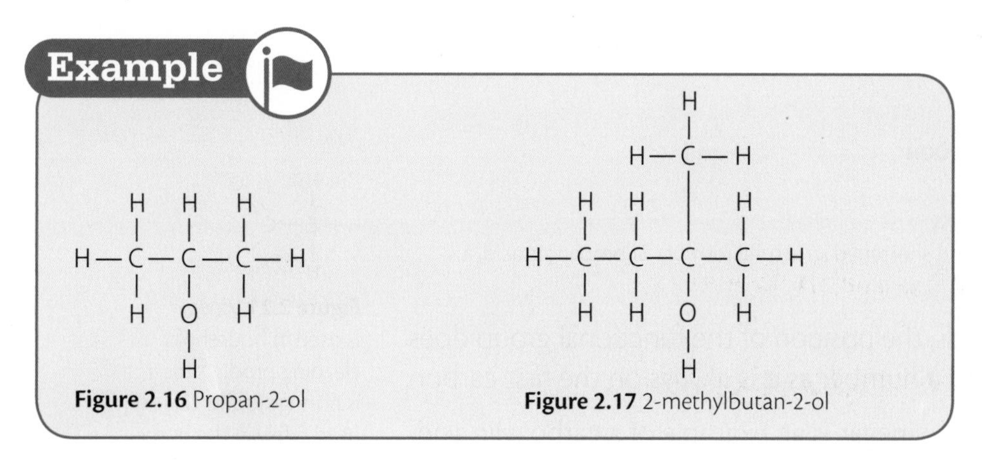

Figure 2.16 Propan-2-ol

Figure 2.17 2-methylbutan-2-ol

Note that in Figure 2.18 the **hydroxyl group** is drawn so that the O of the OH group is bonding to the neighbouring carbon atom. This is so we can see it is the oxygen that is bonding with the carbon, not the hydrogen.

```
     H   H              H   H
     |   |              |   |
HO — C — C — H      H — C — C — OH
     |   |              |   |
     H   H              H   H
```

Figure 2.18 The hydroxyl group in ethanol

Alcohols are a very important family of compounds because of their many and varied uses: as solvents in perfumes, as a replacement for the toxic metal mercury in thermometers and as fuels, because they burn with a clean flame.

Figure 2.19 Alcohols have many uses, including in perfumes and as fuels

As the size of a molecule increases, so does the strength of the intermolecular bonds; this results in the larger alcohol molecules having higher melting and boiling points than smaller molecules.

The smaller alcohol molecules (methanol, ethanol and propanol) are all **miscible** with water, meaning they can be easily mixed with water. The larger alcohols are not miscible with water and form two layers when mixed with water.

Key point

* As alcohol molecules increase in size, their melting and boiling points increase due to the increasing strength of the intermolecular bonds.

Carboxylic acids

Carboxylic acids are a family of compounds with the **carboxyl** functional group (–COOH).

```
    H    H    O
    |    |    ||
H — C —  C —  C
    |    |     \
    H    H      OH
```

CH_3CH_2COOH

Figure 2.20 The structural formula and shortened structural formula of propanoic acid, a member of the carboxylic acid family

When naming carboxylic acids, the position of the functional group does not have to be indicated with a number as it is *always* on the first carbon.

Ethanoic acid, better known as vinegar, is an example of a carboxylic acid.

```
    H     O
    |    //
H — C — C
    |    \
    H     O — H
```

Figure 2.21 Vinegar is used in household cleaning products as it is non-toxic and can be used safely in the home

Figure 2.22 Good job these chemicals are non-toxic!

Carboxylic acids are used in the manufacture of esters, soap and medicines such as aspirin.

The smaller carboxylic acid molecules (methanoic acid, ethanoic acid and propanoic acid) are all miscible with water, meaning they can be easily mixed with water. The larger carboxylic acids are not miscible with water and form two layers when mixed with water.

As the size of a molecule increases, so does the strength of the intermolecular bonds; this results in the larger carboxylic acid molecules having higher melting and boiling points than smaller molecules.

Since they are similar to other acids, carboxylic acids react in neutralisation reactions with metal oxides, metal hydroxides, metal carbonates to form salts.

Key point 

* As carboxylic acid molecules increase in size their melting and boiling points increase due to the increasing strength of the intermolecular bonds.

Remember

It is important to be able to name the salts produced in the neutralisation reactions of carboxylic acids.

Acid name	Name of salt ends in
methanoic acid	...methanoate
ethanoic acid	...ethanoate
propanoic acid	...propanoate

Example 

1 methanoic acid + sodium hydroxide → sodium methanoate + water

2 ethanoic acid + calcium carbonate → calcium ethanoate + water + carbon dioxide

3 propanoic acid + magnesium oxide → magnesium propanoate + water

Study questions

1 Identify the correct systematic name for the compound with this shortened structural formula: $CH_3CH(OH)CH_2CH_2CH_3$?

a) pentane
b) pentan-2-ol
c) pentan-3-ol
d) 2-methylpentane

2 Which of the following can be classed as a carboxylic acid?

a) $C_6H_{14}O$
b) $C_6H_{13}OH$
c) $C_6H_{13}COOH$
d) $C_2H_5COOC_3H_7$

3 Most sterilising pads used in hospitals contain a 65% solution of isopropyl alcohol in water. Isopropyl alcohol has the systematic name propan-2-ol (C_3H_7OH).

a) Draw the full structural formula of isopropyl alcohol.
b) Name an isomer of propan-2-ol.
c) A typical sterilising pad contains approximately 0·6 g of propan-2-ol. Calculate the number of moles of propan-2-ol that a typical sterilising pad contains.

4 Formic acid (methanoic acid) is contained in the stings of many insects, such as wood ants, and is also present in stinging nettles.

a) Draw the full structural formula of formic acid.
b) State the name of the functional group contained in formic acid.
c) Formic acid can be reacted with butan-1-ol to produce a compound called an ester. Draw the full structural formula of butan-1-ol.

Chapter 2.3
Combustion

Fuels are vitally important to humans. They provide the energy we need to live, such as for heating our homes and powering our transport. Combustion releases the energy that is contained in fuels.

When a substance burns, it reacts with oxygen in a reaction called combustion. All combustion reactions are **exothermic** because they *release* energy, usually in the form of heat. For example, when methane is burned using a Bunsen burner, heat energy is given out.

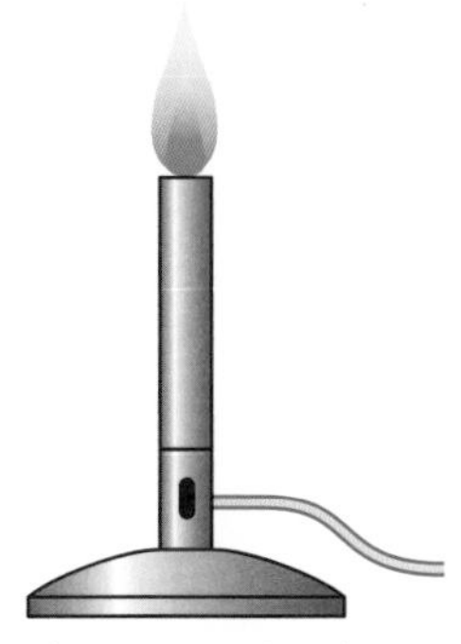

Figure 2.23 Combustion of methane using a Bunsen burner

The opposite of an exothermic reaction is an **endothermic** reaction. An endothermic reaction *takes in* energy, resulting in a drop in temperature.

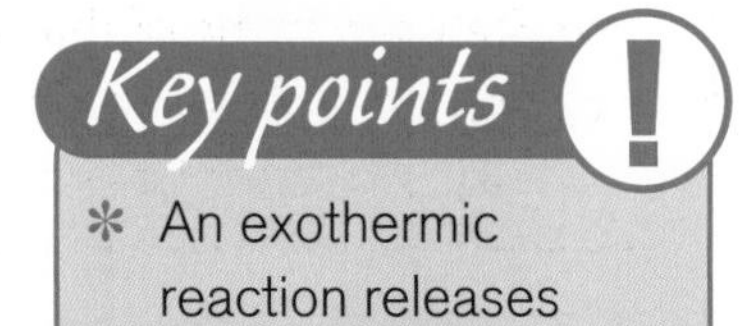

* An exothermic reaction releases heat energy.
* An endothermic reaction takes in energy.

Figure 2.24 Instant ice packs make use of an endothermic reaction

For combustion to take place, a fuel is required. Alkanes and alcohols are commonly used fuels.

Alkanes, such as methane gas, are used to heat your home and cook your food.

Alcohols, such as ethanol, can be used to power cars; in Brazil, ethanol is more commonly used than petrol to power 'FlexFuel' cars, These cars can run on petrol, or ethanol, or a mixture of both.

Figure 2.25 FlexFuel cars run on petrol and/or ethanol

All hydrocarbons and alcohols burn in a plentiful supply of oxygen, producing carbon dioxide and water.

In combustion, the oxygen combines with each individual element in the compound. The example below shows what happens when methane (CH_4) is burned. The oxygen combines with the carbon and the hydrogen in the methane molecule to produce carbon dioxide and water.

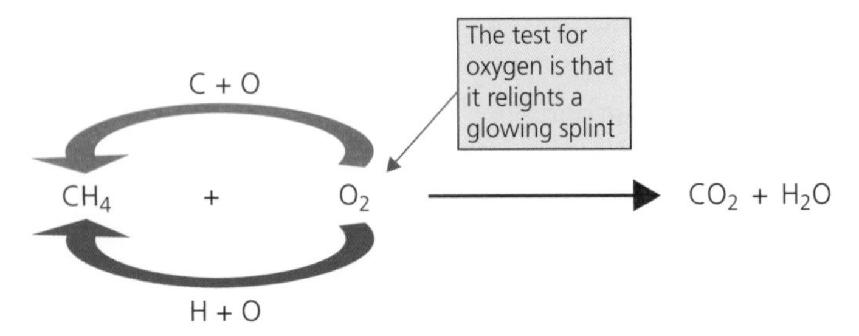

Figure 2.26

When any hydrocarbon is burned in a plentiful supply of oxygen, carbon dioxide (CO_2) and water (H_2O) are produced.

The products of combustion can be identified by performing the experiment shown in Figure 2.27.

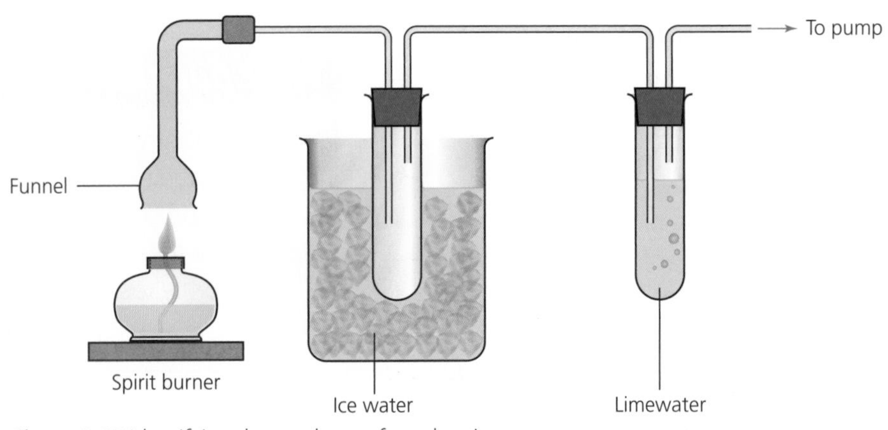

Figure 2.27 Identifying the products of combustion

To test for carbon dioxide, limewater is used. If carbon dioxide is present, the limewater turns from colourless to milky white.
If water is produced, it will condense in the test tube in the ice-water bath.

If the volume of air is restricted, then incomplete combustion takes place. Due to the lack of oxygen, the toxic gas carbon monoxide is produced.

Chapter 2.4 Energy from fuels

When a chemical reaction takes place, there is usually a change in energy. This difference in energy between the reactants and products can be calculated.

Calculating energy

The change in energy can be calculated from the experimental data obtained by performing the experiment shown in Figure 2.28.

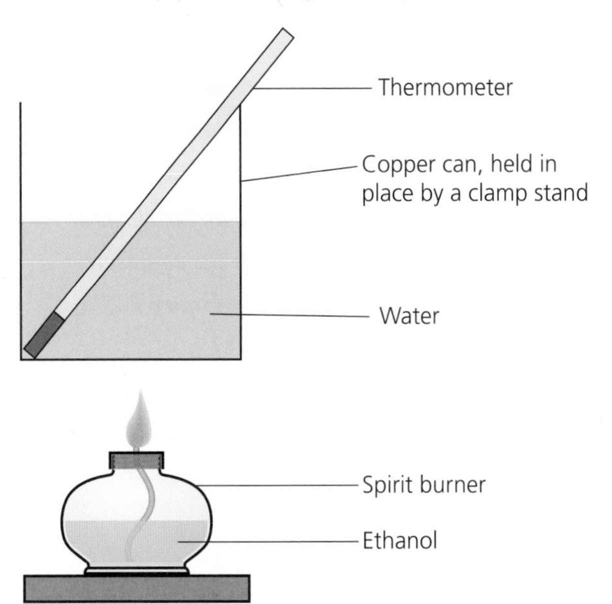

Figure 2.28 Apparatus used in the lab for establishing the energy released from the combustion of ethanol

These results can then be added to the equation shown in Figure 2.29 to calculate the change in energy. A verson of this equation can be found on page 3 of the Data Booklet.

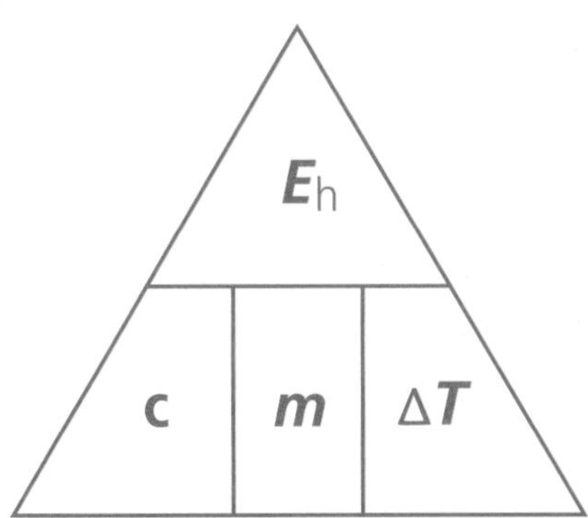

E_h - is the enthalpy change (kJ)
c - is the specific heat capacity of water. It is a constant, 4·18 kJ kg^{-1} °C^{-1}
m - is the mass of water (in kg) (100 cm^3 = 0·1 kg) (1 l = 1 kg) **Do not use the mass of fuel here!**
ΔT - is the **change** in temperature (°C)

Figure 2.29 Enthalpy equation triangle

Example 1

The following data were recorded when ethanol (C_2H_5OH) was burned. Use the information in the table to calculate the enthalpy change, in kJ, for this reaction.

Data measured	Result
mass of burner and ethanol before	53·01 g
mass of burner and ethanol after	52·65 g
mass of water heated	200 cm^3
initial temperature of water	21°C
final temperature of water	26°C

temperature change (ΔT) = 26 − 21 = 5°C

mass of water used = $\frac{200}{1000}$ = 0·2 kg

$$E_h = cm\Delta T$$

$$= 4{\cdot}18 \times 0{\cdot}2 \times 5$$

$$= 4{\cdot}18\ \text{kJ}$$

Note that the mass of ethanol used is not required here; don't get it confused with the mass of water.

Hints & tips

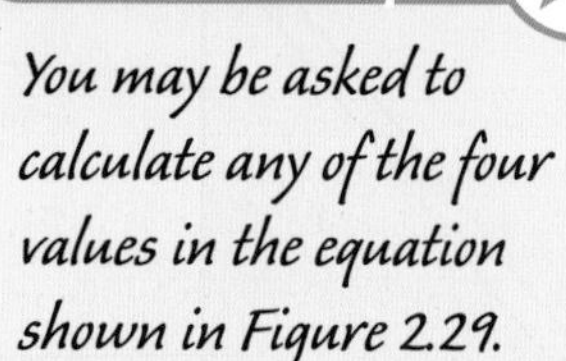

You may be asked to calculate any of the four values in the equation shown in Figure 2.29.

$$E_h = cm\Delta T$$

$$\Delta T = \frac{E_h}{cm}$$

$$c = \frac{E_h}{\Delta T m}$$

$$m = \frac{E_h}{c\Delta T}$$

Example 2

Burning 3 g of methanol releases 70 kJ of energy. If 1 l of water is heated using this energy, calculate the change in temperature, in °C.

$$\Delta T = \frac{E_h}{cm}$$

$$= \frac{70}{4{\cdot}18 \times 1}$$

$$= 16{\cdot}75\text{°C}$$

Study questions

1 Oxygen gas is mixed with methane gas and ignited. What is the correct balanced equation for this reaction?

a) $CH_4 + 2O_2 \rightarrow CO_2 + 2H_2O$
b) $CH_4 + O_2 \rightarrow CO_2 + 2H_2O$
c) $CH_4 + 2O_2 \rightarrow 2CO_2 + 2H_2O$
d) $2CH_4 + 2O_2 \rightarrow 2CO_2 + 2H_2O$

2 The balanced equation for the complete combustion of a hydrocarbon **X** is shown below.

Which of the following options is the correct formula of hydrocarbon **X**?

$$\mathbf{X} + 8O_2 \rightarrow 5CO_2 + 6H_2O$$

a) C_2H_6
b) C_3H_8
c) C_5H_{14}
d) C_5H_{12}

3 Many chemical reactions involve energy changes and the energy that is released can be put to use.

a) Hand warmers make use of an increase in temperature that takes place during a reaction within the warmer. State the name of the type of reaction that results in an increase in temperature.

b) Flameless ration heaters are used by the armed forces to prepare meals when on missions. The energy released heats 330 g of water by 37·8°C.
Calculate the energy, in kJ, required to do this.

4 When 1 mole of propan-1-ol is burned, 2021 kJ of energy is released.
If this energy was used to heat 6 kg of water, calculate the temperature change of the water in °C.

Section 3 Chemistry in society

Chapter 3.1 Metals

Over three-quarters of the known elements are metals – they appear on the left-hand side of the 'zig-zag' line on the Periodic Table (Figure 1.16). In this chapter we will revise some reactions that involve metals, and we will start to look at why metals conduct electricity.

Metallic bonding

Metallic bonds occur between the atoms of metal elements. The outer electrons are delocalised (free to move). An electrostatic force of attraction forms between the positive metal ions (atoms that have lost an electron to become positively charged) and the negative delocalised electrons. This electrostatic force acts as a glue to hold the atoms together.

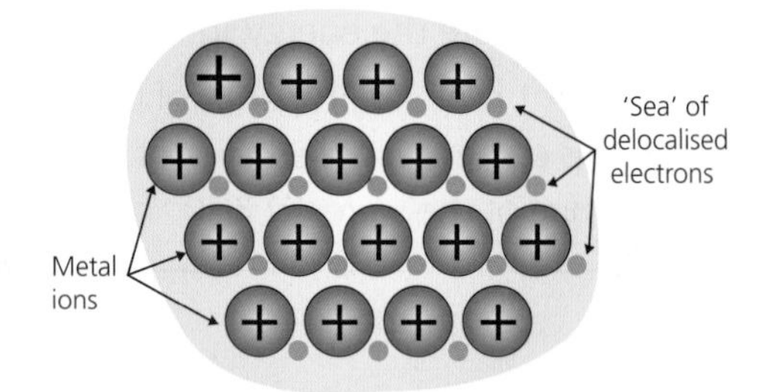

Figure 3.1 Metallic bonding

The delocalised electrons are the reason that metals conduct electricity. Electricity is a flow of electrons. For a substance to conduct electricity, it must allow these electrons to flow through it and metals do this very well.

As one electron moves into the metal from the power supply, one will jump off the metal.

During electrical conduction, the metal remains completely unchanged as it always has the same amount of electrons.

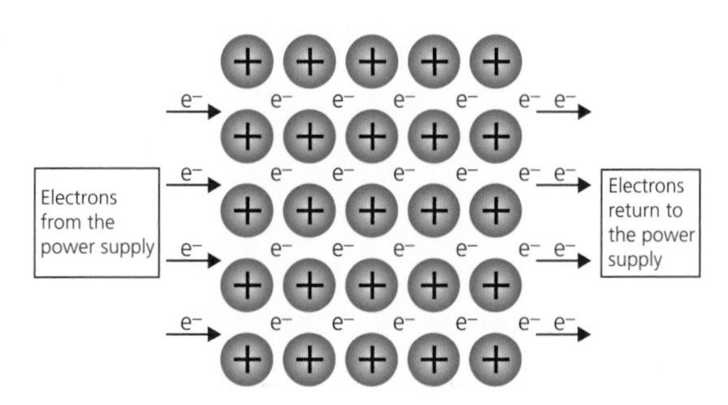

Figure 3.2 The conduction of electricity by a metal

Chapter 3.2
Reactions of metals

Metals and water

Most metals don't react with water, but some of the more reactive metals do. For example, the alkali metals, lithium, sodium and potassium, react very strongly, and caesium can even explode in water!

Key point

reactive metal + water → metal hydroxide + hydrogen

The test for hydrogen gas is that it burns with a pop!

Example

The reaction of lithium and water to produce lithium hydroxide and hydrogen gas can be summarised as follows:

Formula equation:

$$2Li(s) + 2H_2O(l) \rightarrow 2LiOH(aq) + H_2(g)$$

Ionic equation:

$$2Li(s) + 2H_2O(l) \rightarrow 2Li^+(aq)\ 2OH^-(aq) + H_2(g)$$

Hints & tips

Gold, silver and platinum don't react with water at all, because they are unreactive metals. The ***reactivity series*** *is on page 71.*

Reactions with acids

As with water, it is just the more reactive metals that react with acids. Only the metals above hydrogen in the reactivity series react with acids, because all acids contain hydrogen (H^+) ions. Acids do not have an effect on metals such as gold and platinum.

Key point

metal + acid → salt + hydrogen

Example 

The reaction of magnesium and hydrochloric acid to produce magnesium chloride and hydrogen gas can be summarised as follows:

Formula equation:

$$Mg(s) + 2HCl(aq) \rightarrow MgCl_2(aq) + H_2(g)$$

Ionic equation:

$$Mg(s) + 2H^+(aq)\ 2Cl^-(aq) \rightarrow Mg^{2+}(aq)\ (Cl^-)_2(aq) + H_2(g)$$

Note that, in this reaction, Cl^- is a spectator ion.

The reaction of metals with acids can be used to produce soluble salts. An excess of metal is added to an acid and the resulting mixture is filtered to remove the metal that did not react. The filtrate is evaporated to dryness, leaving the soluble salt.

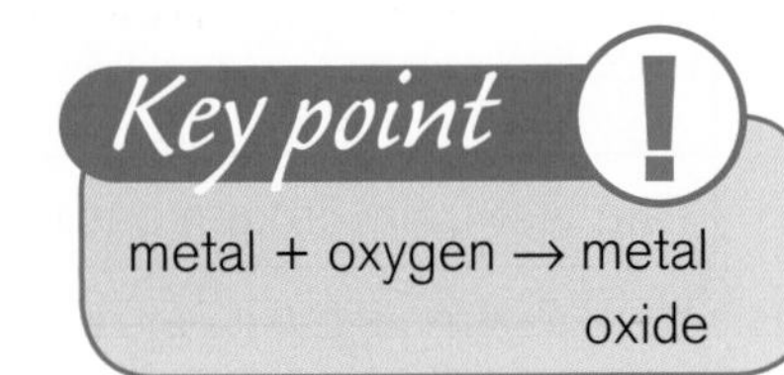

Reactions with oxygen

Again, only the most reactive metals react strongly with oxygen.

Example 

In class, you may have burned magnesium. This reacts very strongly with oxygen to give off a bright light in a very exothermic reaction.

$$2Mg(s) + O_2(g) \rightarrow 2MgO(s)$$

Metal oxides are classed as bases because they neutralise acids (see page 40).

Electrochemical cells

The reactions of metals make them very useful; one example of this is their use in batteries. Batteries provide electricity for our use; in chemistry these are known as **electrochemical cells**.

To understand how an electrochemical cell works, we must know what electricity is. Electricity is a flow of electrons that travels along a wire.

In a cell, this flow of electrons or electricity is produced by a chemical reaction that is taking place in the cell.

Cells have one major disadvantage: they eventually run out. This happens when the chemicals in the cell are used up.

Before we revise in detail how an electrochemical cell works, we must remind ourselves of the reactivity series, which is a list of elements arranged in order of reactivity, starting with the most reactive. It is very similar to the electrochemical series in your Data Booklet, but with a few differences.

The reactivity series is easy to remember if you learn the mnemonic in Figure 3.3.

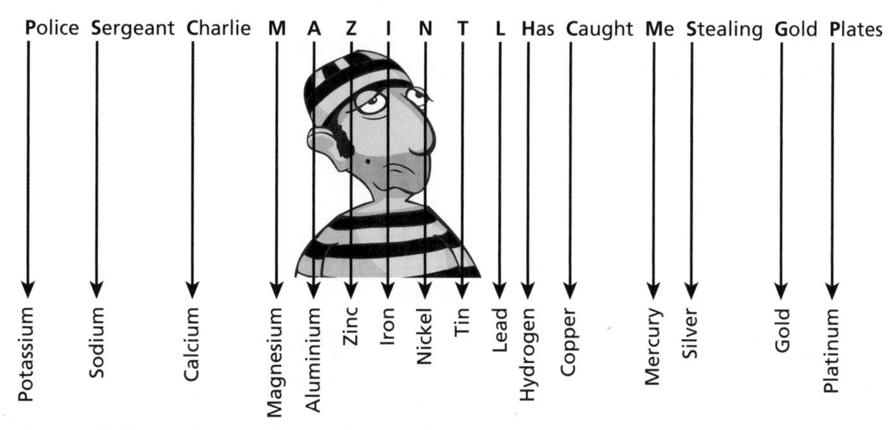

Figure 3.3 Remembering the reactivity series

Cell story

This story is completely fictional but if you learn it, this chapter will become a lot easier.

Magnesium and Copper are cowboys in the old Wild West. To win the heart of an admirer, Mg challenges Cu to a 'shoot out' at high noon. They meet at the appointed time, stand back to back, walk 12 paces, turn and shoot – BANG! Mg shoots Cu with a bullet and wins the duel. Magnesium was the fastest to REACT.

This story may seem out of place in a chemistry book but all will become clear when you read on. The bullet in the story stands for electrons. So, as the bullet flies from Magnesium's gun, electrons flow from the magnesium (faster to react) to the copper (slower to react).

The electrons always go from the *most reactive metal* to the *least reactive metal.*

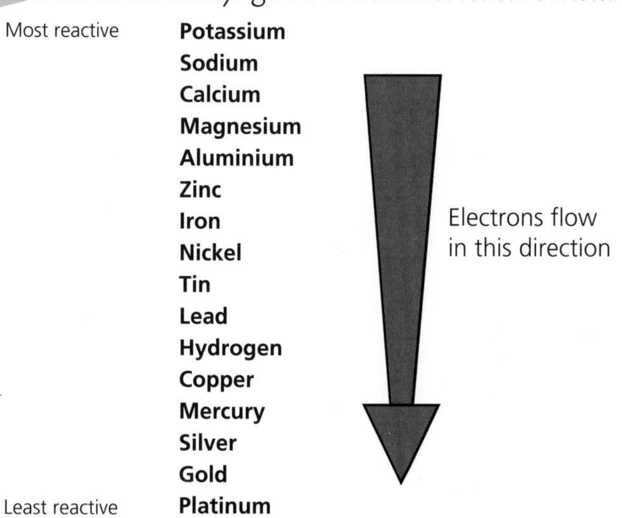

Figure 3.4 The reactivity series

Electricity is a flow of electrons, therefore electricity has been made when this happens.

Cells

A cell is made up of two different metals connected by an **electrolyte** (to complete the circuit). An electrolyte is an ionic solution that conducts electricity.

The flow of electrons is produced because of the difference in reactivity of the two metals.

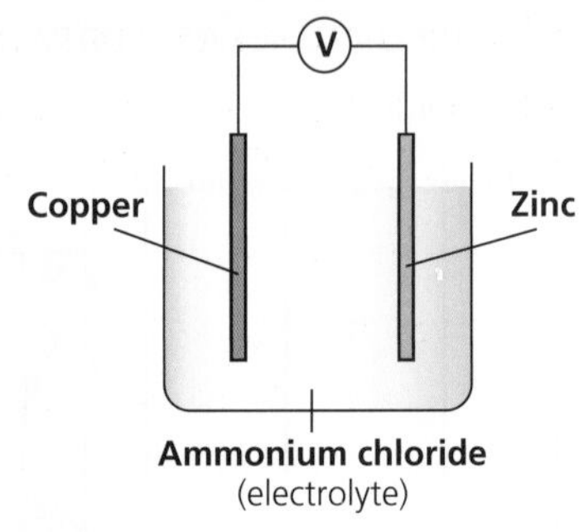

Figure 3.5 An example of a cell

Key points

* The flow of electrons in a cell is due to the difference in reactivity of the two metals.
* Electrons will flow from the most reactive metal to the least reactive metal.
* An electrolyte is an ionic solution that is used to complete the circuit.

Hints & tips

Read page 22 to remind yourself why an electrolyte has to be an ionic solution. Remember the story on page 71! In the cell, the electrons will flow from the zinc to the copper, because zinc is faster to react than copper.

So, in the cell shown in Figure 3.6 the electrons flow from the zinc to the copper, because zinc is more reactive than copper.

The further apart the metals are in the electrochemical series, the higher the voltage produced. If we replaced the zinc in the cell in Figure 3.6 with magnesium, the voltage produced by the cell would increase, because magnesium is even more reactive than zinc.

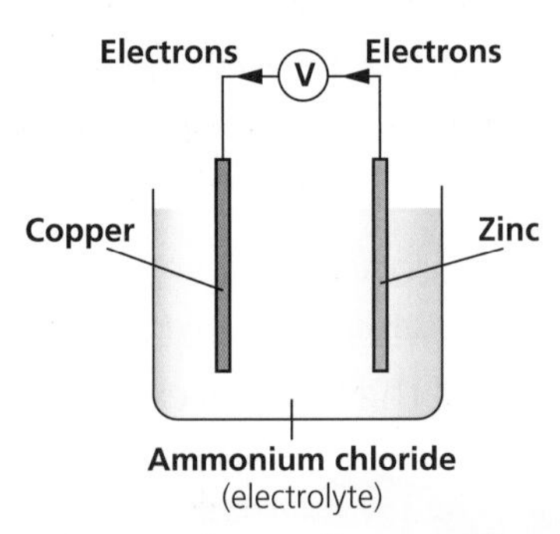

Figure 3.6 Electron flow in a cell

Key point

* The further apart the metals are in the electrochemical series, the larger the voltage produced by the cell.

Different cells

There are several different ways in which cells can be made.

Electricity can also be produced in a cell by connecting two different metals in a solution of their own ions.

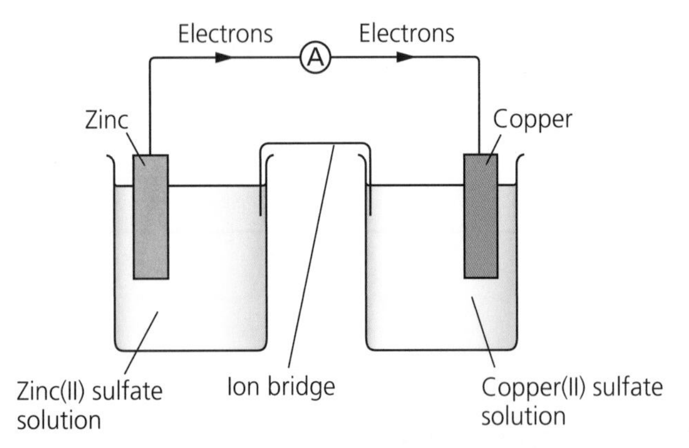

Figure 3.7 An example of a cell – the ion bridge completes the circuit

In Figure 3.7, the electrons will flow from the zinc to the copper, as in the cell in Figure 3.6. This will cause the zinc to dissolve and more copper to form on the surface of the copper electrode.

As with any electrical circuit, it must be complete to work. The purpose of the **ion bridge** is to complete the circuit. The ion bridge is not some piece of high-tech electrical equipment! It is simply a piece of filter paper soaked in salt water or some other ionic solution. Ions can move across the bridge to complete the circuit (Figure 3.8).

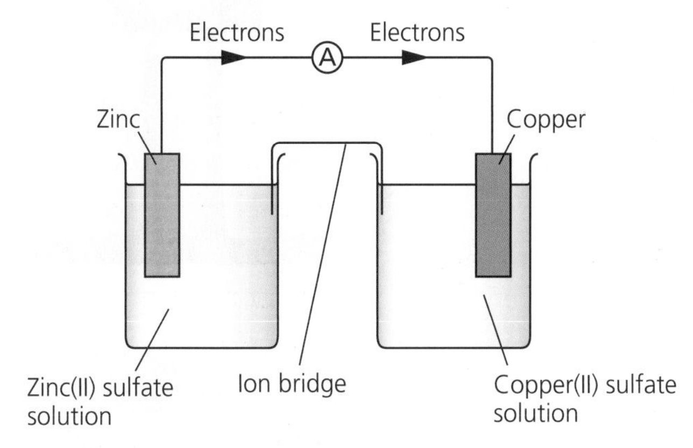

Figure 3.8 Electron flow in a cell with an ion bridge

Another type of cell is one in which only one metal is involved and the other half of the cell contains SO_4^{2-} ions or I_2 with a graphite electrode. As in the cells we looked at previously, the electrons will travel from the substance that is higher in the electrochemical series to the substance that is lower down in the series (Figure 3.9).

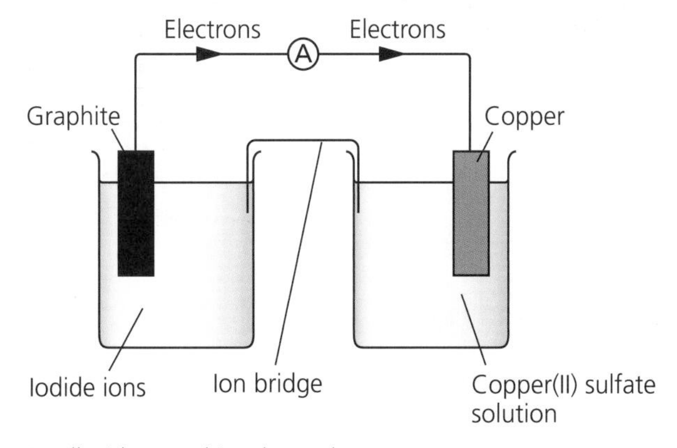

Figure 3.9 A cell with a graphite electrode

Oxidation and reduction

Substances can gain or lose electrons.

* When substances lose electrons, this is oxidation.
* When substances gain electrons, this is reduction.

When a metal element is reacting to form a compound, it is being oxidised. In the following equation, the magnesium is losing two electrons, so this is an **oxidation reaction**.

$$Mg \rightarrow Mg^{2+} + 2e^-$$

A **reduction reaction** is the opposite of oxidation. It is the gain of electrons. For example,

$$Cu^{2+} + 2e^- \rightarrow Cu$$

A **redox reaction** is one in which both oxidation and reduction take place.

Hints & tips

There is an easy way to remember oxidation and reduction using an oil rig.

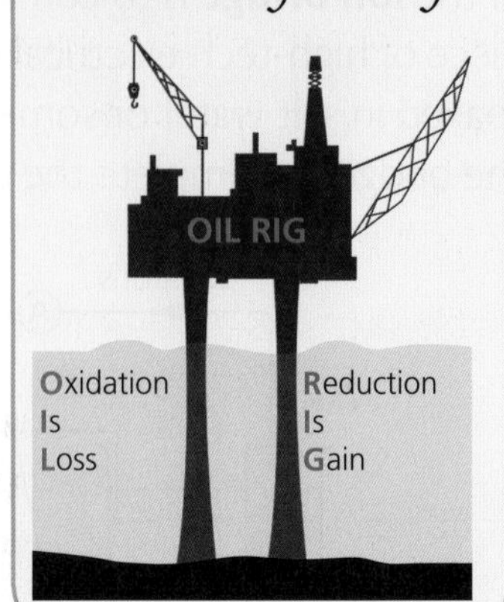

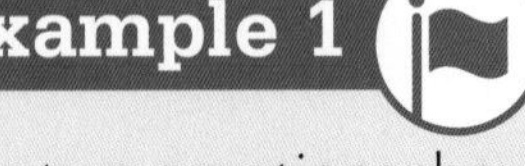

Example 1

The two equations shown above and here can be combined to form a redox equation.

Oxidation:

$$Mg \rightarrow Mg^{2+} + 2e^-$$

Reduction:

$$Cu^{2+} + 2e^- \rightarrow Cu$$

Redox:

$$Mg + Cu^{2+} \rightarrow Mg^{2+} + Cu$$

Hints & tips

Page 10 of your Data Booklet lists the reduction reactions of substances in the electrochemical series.

To combine the two equations to form a redox equation, both equations must show an equal number of electrons, so that the loss and gain cancel out. To achieve this, one or both of the equations may have to be multiplied.

Example 2

This equation must be multiplied by two so that there is an equal number of electrons in both equations.

Reduction:

$$Ag^+ + e^- \rightarrow Ag$$

Oxidation:

$$Mg \rightarrow Mg^{2+} + 2e^-$$

Multiply the Ag equation throughout by two:

$$2Ag^+ + 2e^- \rightarrow 2Ag$$

Now both equations have the same number of elections:

$$Mg \rightarrow Mg^{2+} + 2e^- \quad 2Ag^+ + 2e^- \rightarrow 2Ag$$

Combine the equations for a redox reaction:

$$2Ag^+ + Mg \rightarrow Mg^{2+} + 2Ag$$

Extraction of metals

When metals are found in nature, they have been present for a very long time. Usually, over many years, they have reacted with the oxygen in the air and water. So, most of them are compounds not pure metals. Naturally occurring impure metal compounds are called **ores**.

The less reactive metals, such as gold and platinum, don't form compounds because they are so unreactive. However, they may be mixed with impurities.

To be useful, most metals have to be removed from their ores. The more reactive the metal, the more difficult it is to remove from its ore.

Key points 

* Unreactive metals are removed from their ores simply by heating.
* Some metal ores need to be heated with carbon (or carbon monoxide) to remove the oxygen from metal compounds in the ore.
* The most reactive metals require **electrolysis** to remove the impurities.
* The table below shows some metals and the processes that are used to extract them from their ores.

Metal	Process by which metal is extracted from ore
potassium sodium lithium calcium magnesium aluminium	electrolysis
zinc iron tin lead copper	heat plus carbon or carbon monoxide
mercury silver gold	heat alone

Electrolysis

Only ionic solutions can be used in electrolysis as covalent compounds generally don't conduct electricity (page 24).

Key point 

* Electrolysis is the breaking up of an ionic solution using electricity.

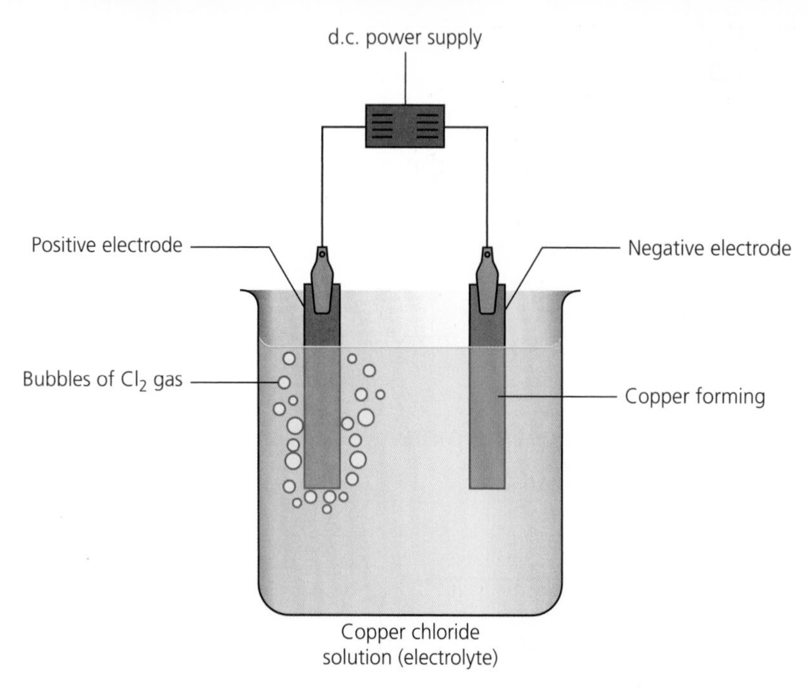

Figure 3.10 Electrolysis of copper chloride

In Figure 3.10, the copper chloride solution is broken up. When electricity is passed through the solution, the positive copper ions are attracted to the negative electrode.

When the copper ions get to the electrode, they pick up two electrons to form copper metal in a reduction reaction:

$$Cu^{2+}(aq) + 2e^{-} \rightarrow Cu(s)$$

The negatively charged chloride ions are attracted to the positive electrode. When they get there, they lose their extra electrons to become chlorine gas in an oxidation reaction:

$$2Cl^{-}(aq) \rightarrow Cl_2(g) + 2e^{-}$$

During electrolysis a **direct current (d.c.) power supply** must be used so that the products can be identified.

Hints & tips

Chlorine gas is a diatomic molecule. That is why the equation for the formation of chlorine by oxidation involves two electrons, not one as you might expect.

Study questions

1 Metallic bonds are due to
   a) a shared pair of electrons
   b) an attraction between positive ions and negative ions
   c) an attraction between positive ions and delocalised electrons
   d) an attraction between negative ions and delocalised electrons.

2 $Fe^{2+}(aq) + e^{-} \rightarrow Fe^{+}(aq)$

   This ion–electron equation represents the
   a) reduction of iron(II) ions
   b) reduction of iron(I) ions
   c) oxidation of iron(II) ions
   d) oxidation of iron(I) ions.

3 Which pair of metals will produce an electron flow in the direction shown in Figure 3.11?
   a) Mg/Zn
   b) Zn/Mg
   c) Fe/Zn
   d) Au/Zn

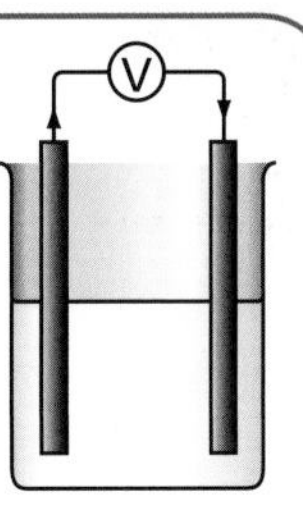

Figure 3.11

4 Which of the following metals can be extracted from its ore by heat alone?
   a) aluminium
   b) zinc
   c) gold
   d) iron

5 Experiments were performed on three unknown metal elements, X, Y and Z to try to establish their reactivity. The results of the experiments are recorded in the table below.

Metal	Reaction with water	Reaction with dilute acid
X	No reaction	No reaction
Y	Slow reaction	Fast reaction
Z	No reaction	Slow reaction

   The order of reactivity of the metals, starting with the **least** reactive, is
   a) Z, X, Y
   b) X, Z, Y
   c) Y, Z, X
   d) X, Y, Z

6 All metals can conduct electricity because
   a) the electricity breaks the bonds between the atoms
   b) the delocalised electrons in metals are free to move
   c) the shared electrons between the atoms attract stray protons
   d) the metals contain mobile ions.

7 In Figure 3.12 what is the purpose of the ion bridge?

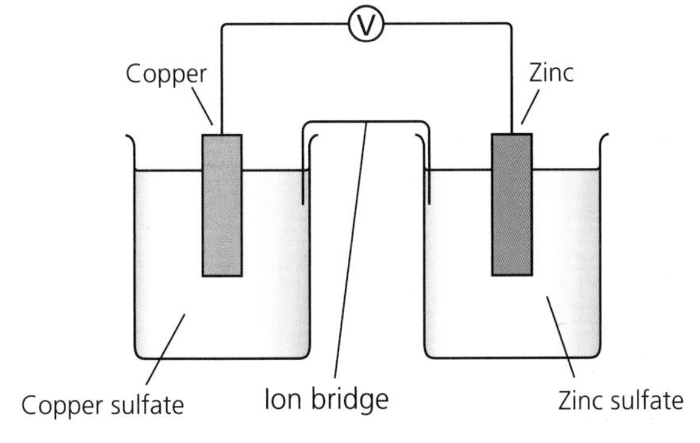

Figure 3.12

   a) to complete the circuit
   b) to provide ions
   c) to indicate that a reaction has taken place
   d) to act as a catalyst

8 Which of the following metals would react with dilute hydrochloric acid to produce hydrogen gas?

a) silver
b) calcium
c) mercury
d) gold

9 A student investigated the effect of changing one of the metal electrodes in a cell on the voltage produced when using sodium chloride as an electrolyte.

a) Draw and label a diagram of the equipment needed to measure the voltage produced when zinc and iron are used as electrodes.

b) The results obtained by the student were recorded in the table.

Use the results obtained to predict the voltage that would be produced when a tin electrode is used with zinc, and predict the direction of electron flow when an aluminium electrode is used with zinc.

Electrodes	Reading on voltmeter (V)	Direction of electron flow
Zn/Fe	0·3	Zn → Fe
Zn/Al	−0·1	
Zn/Ni	0·5	Zn → Ni
Zn/Sn		Zn → Sn

10 An experiment to list the metals copper, magnesium and zinc in order of reactivity was performed by reacting each of the metals with oxygen gas.

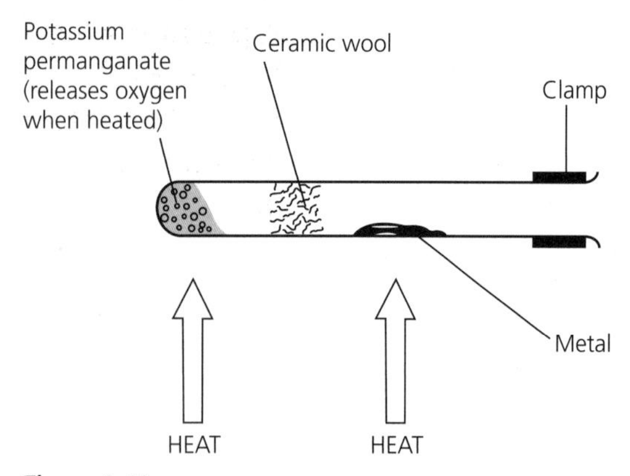

Figure 3.13

a) What is the purpose of the potassium permanganate?

b) List the metals in order of reactivity from the most to the least reactive.

11 When copper is added to a solution of silver nitrate the solution turns blue and the copper turns from brown to silver.

$$2AgNO_3(aq) + Cu(s) \rightarrow Cu(NO_3)_2(aq) + 2Ag(s)$$

a) Name the spectator ion in this reaction.

b) Write the ion–electron equation for the oxidation step in this reaction.

Chapter 3.3

Properties of plastics

There are many different types of plastic, all with very different properties. This gives them a massive range of uses, from clothing to packaging to children's toys. This chapter highlights the advantages and disadvantages of plastics, as well as looking at how plastics are produced.

Synthetic is the name given to all materials that are made by humans. All plastics are synthetic compounds, made mainly from crude oil.

There are many advantages of plastics over natural materials and some disadvantages. For example, cotton (a natural fibre) is more comfortable to wear, but nylon (synthetic) is more hardwearing.

Plastics are made by a process called **polymerisation.**

Addition polymerisation

Polymerisation is a process by which many small molecules, called monomers, combine to form one large molecule called a polymer (the plastic).

You already know about the addition reactions of alkenes with bromine (page 53). The addition concept is the same in **addition polymerisation**, but on a much larger scale.

Key points

* **Monomers** are the small unsaturated molecules that combine to form the polymer.
* A **polymer** is the large molecule formed by combining many monomer molecules.

Example 1

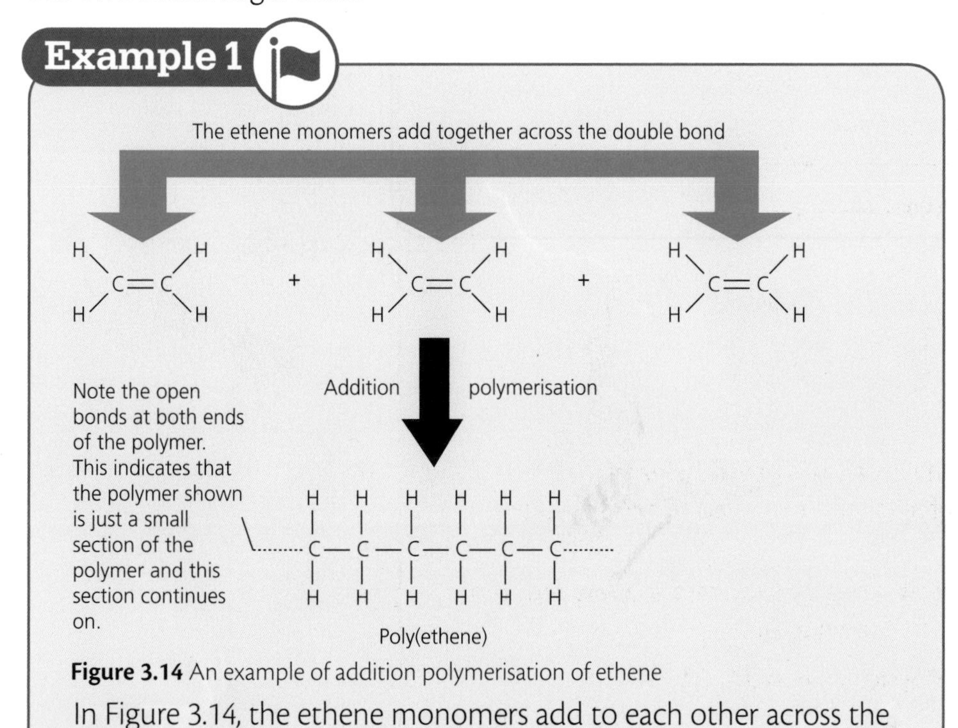

Figure 3.14 An example of addition polymerisation of ethene

In Figure 3.14, the ethene monomers add to each other across the double bond, forming the polymer poly(ethene).

The monomers for addition polymerisation must be unsaturated (contain a carbon-to-carbon double or triple bond) in order for the additions to take place.

Naming the polymer produced by addition polymerisation is very simple – just add 'poly' to the start of the monomer name:

- ethene produces poly(ethene)
- propene produces poly(propene).

Because the polymer molecules produced are very large, chemists usually just draw the repeating unit. This is a shortened version of the polymer molecule.

Example 2

Figure 3.15 shows the repeating unit for poly(ethene).

```
 /    H     H    \
|     |     |     |
| ---C  —  C---   |
|     |     |     |
 \    H     H    / n
```

Where *n* is any large number.

Figure 3.15 Poly(ethene) repeating unit

Example 3

Figure 3.16 shows the repeating unit for poly(propene).

```
 /    H    CH3   \
|     |     |     |
| ---C  —  C---   |
|     |     |     |
 \    H     H    / n
```

Figure 3.16 Poly(propene) repeating unit

Chapter 3.4
Fertilisers

Fertilisers are often required for the growth of agricultural crops. Year on year, the world population is increasing, which leads to a bigger and bigger demand for food production. Fertilisers are needed in order to meet this demand. This chapter highlights what a fertiliser is and how they are produced.

What is a fertiliser?

Fertilisers are substances that replace essential elements, which are required for healthy plant growth.

Fertilisers provide the three main nutrients that plants require to grow well:

- nitrogen (N)
- phosphorus (P)
- potassium (K).

Ammonia (NH_3)

Ammonia is a very important chemical. It is used as a fertiliser itself, but can also be used to make other soluble nitrogen-containing salts, like ammonium nitrate, another fertiliser. In addition, it is used to make explosives, medicines and cleaning products.

ammonia + nitric acid → ammonium nitrate + water

↑ Soluble nitrogen-containing salt

Ammonia is a colourless, alkaline gas with a very unpleasant (pungent) smell.

Ammonia is very soluble in water. The solubility of ammonia can be demonstrated by performing the 'ammonia fountain' experiment, as shown in Figure 3.18.

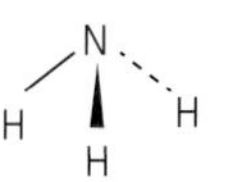

Figure 3.17 Ammonia

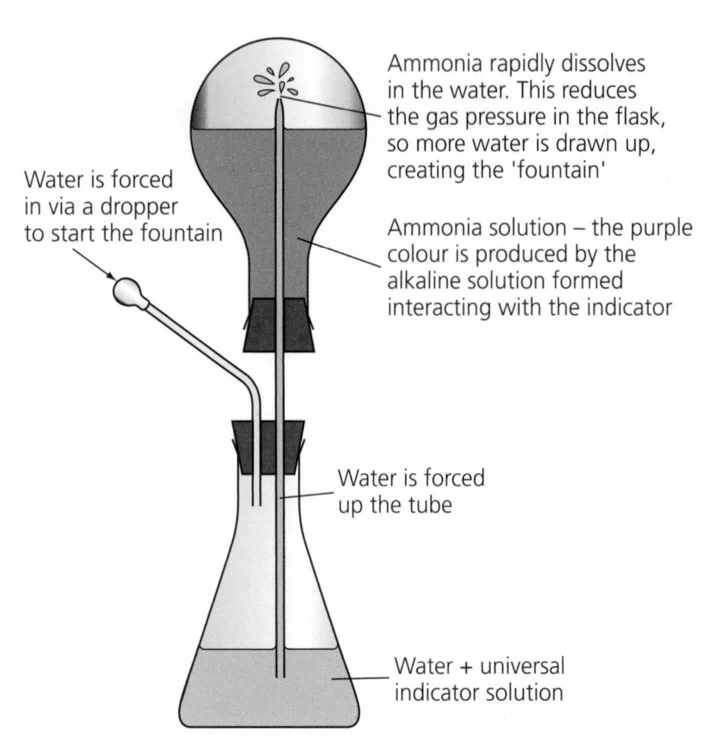

Figure 3.18 The 'ammonia fountain' experiment

Ammonia can be prepared in the lab by heating an ammonium salt with an alkali, such as sodium hydroxide:

$$NH_4Cl(s) + NaOH(aq) \rightarrow NH_3(g) + NaCl(s) + H_2O(l)$$

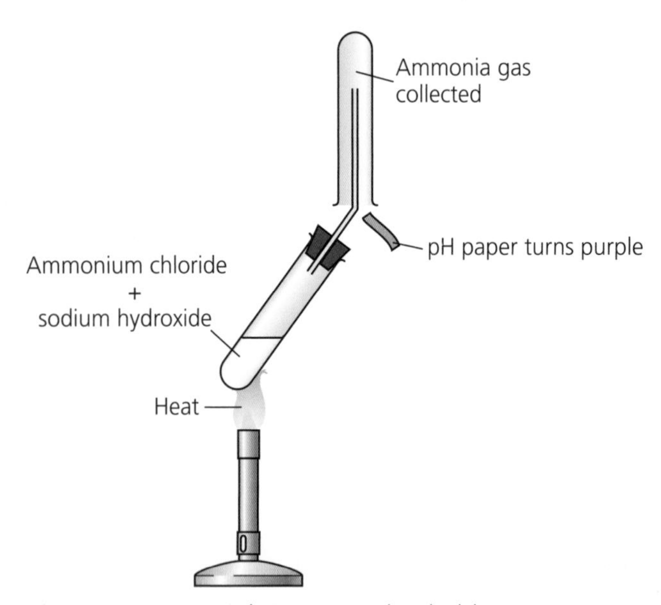

Figure 3.19 Ammonia being prepared in the laboratory

In Figure 3.19 the ammonia gas is being collected by the downward displacement of air. This method can be used to collect any gas that is less dense than air.

In industry, ammonia is produced by the **Haber process**.

The Haber process

The Haber process produces ammonia (NH_3) by combining nitrogen and hydrogen. The nitrogen required comes from the air and the hydrogen comes from steam or methane. They are converted into ammonia by passing them over an iron **catalyst** at a moderately high temperature of 400°C.

These conditions are chosen to maximise the rate of ammonia production. The reaction is faster at a very high temperature but, because the reaction is reversible, the ammonia breaks down back into hydrogen and nitrogen. A moderately high temperature is chosen so that the forward reaction is relatively quick, but the rate of the reverse reaction is not too great.

$$N_2(g) + 3H_2(g) \rightleftharpoons 2NH_3(g)$$

The iron catalyst speeds up the rate of the reaction and is broken down into small pieces to increase its surface area and, therefore, its effectiveness as a catalyst.

The yield of ammonia in the Haber process is only approximately 15%, so all the nitrogen and hydrogen that are not converted into ammonia are put back into the reaction chamber to make the whole process more economic. This recycling of starting materials can be seen in Figure 3.20.

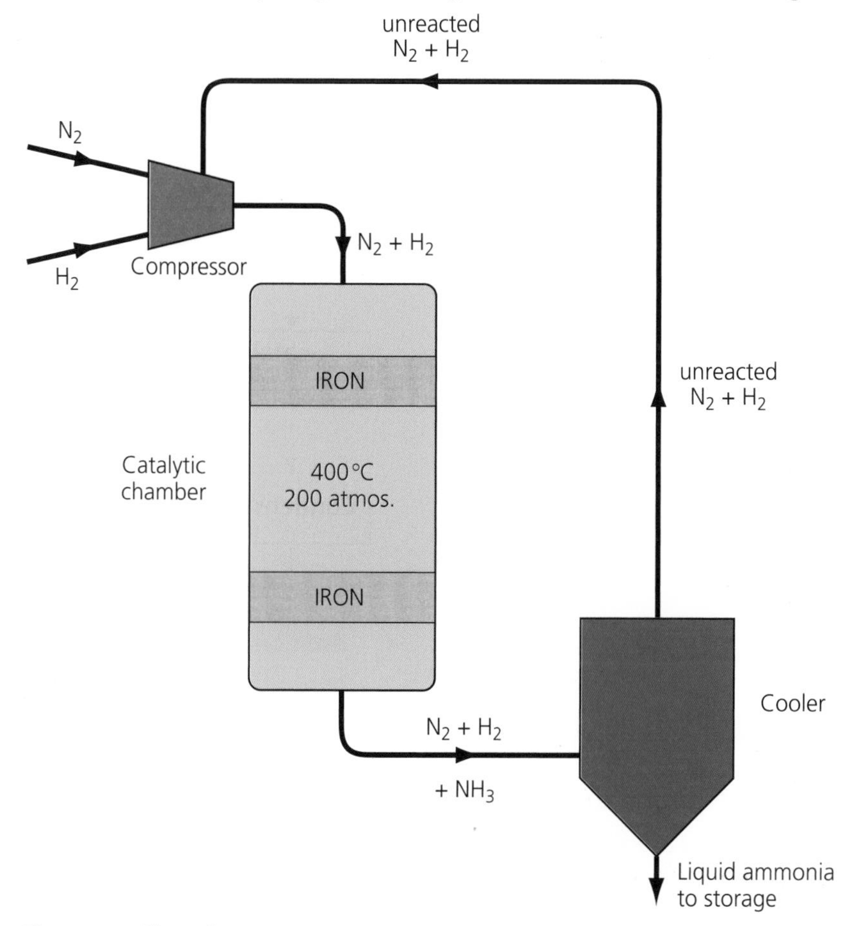

Figure 3.20 The Haber process

The ammonia produced by the Haber process can be used for the commercial production of nitric acid in the Ostwald process.

The Ostwald process

The Ostwald process involves passing ammonia from the Haber process and air over a platinum catalyst at a high temperature of 800°C. This produces nitrogen monoxide, which combines with oxygen to form nitrogen dioxide. This can easily be converted into nitric acid by dissolving in water.

The Ostwald process is an exothermic process, so once the reaction has started the heat can be removed and the catalyst will continue to glow red hot.

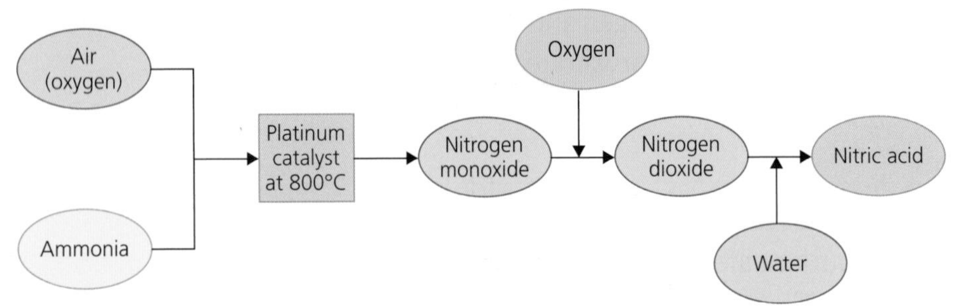

Figure3.21 The Ostwald process

Study questions

1 The Haber process produces ammonia (NH_3) by combining nitrogen and hydrogen.

a) Complete the flow diagram in Figure 3.22 using the following list:

iron nitrogen hydrogen 400 methane

b) Add a line to the flow diagram to show the unreacted hydrogen and nitrogen being recycled.

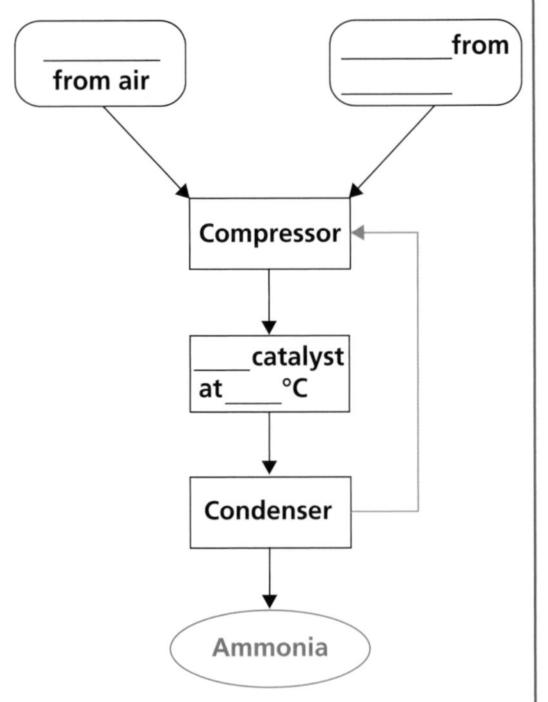

Figure 3.22

2 Styrene, which is also known as phenylethene, can be extracted from the sap of the styrax tree. Styrene is the monomer used to produce polystyrene (Figure 3.23).

Figure 3.23

a) Name the type of polymerisation that takes place to form polystyrene.

b) Draw a section of the polystyrene polymer showing three monomer units combined.

3 Ammonium phosphate, $(NH_4)_3PO_4$, is commonly used by gardeners as a synthetic fertiliser for tomato plants.

a) Calculate the percentage of nitrogen in ammonium phosphate.

b) Calculate the percentage of phosphorus in ammonium phosphate.

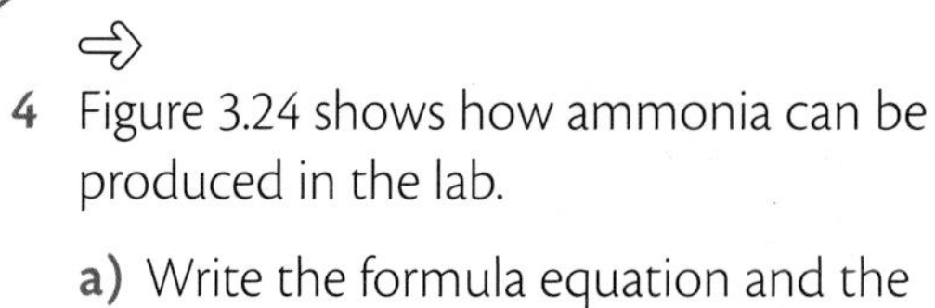

4 Figure 3.24 shows how ammonia can be produced in the lab.

a) Write the formula equation and the ionic equation for this reaction.

b) What does this method suggest about the density of ammonia gas?

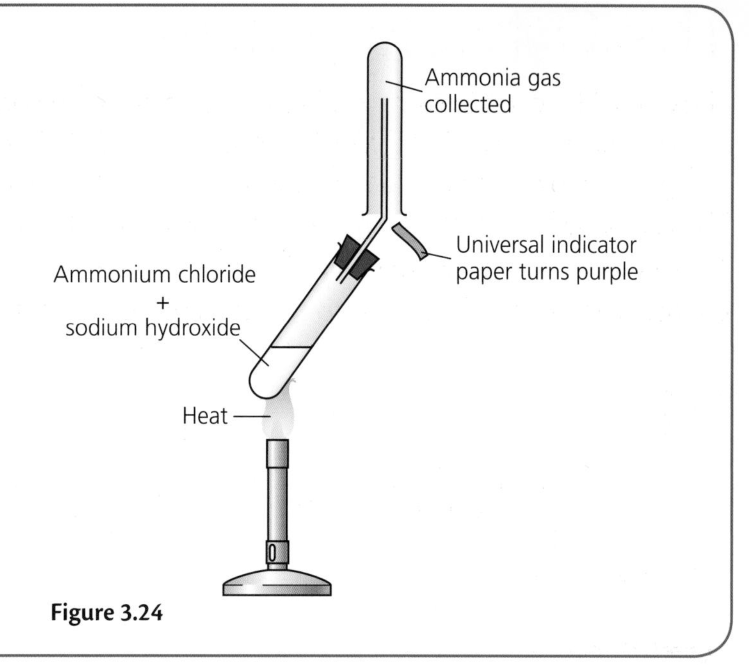

Figure 3.24

Chapter 3.5
Nuclear chemistry

In 1896, French scientist Henri Becquerel discovered that compounds of uranium had an effect on photographic plates from a distance. The radiation they emitted could also penetrate opaque materials. The phenomenon was called radioactivity. In this chapter we will revise the three types of radiation, how to calculate the half-life of a radioactive element, and the different uses of radioisotopes.

Radioactivity

Radioactivity is the result of unstable nuclei emitting energy or a particle to form more stable nuclei.

There are three types of radiation:

- alpha (α)
- beta (β)
- gamma (γ)

Each type of radioactivity has different properties.

Alpha particles

$^{4}_{2}He^{2+}$

Alpha particles are slow-moving, positively charged particles that come from the nucleus of a radioactive element. They consist of two protons and two neutrons and have a 2+ charge.

Alpha particles have little penetration and are stopped by a few centimetres of air, or a sheet of paper.

The nuclear equation below shows an example of nuclear decay with the emission of an alpha particle.

$$^{232}_{90}Th \rightarrow {}^{228}_{88}Ra + {}^{4}_{2}He^{2+}$$

Hints & tips

In nuclear equations the mass numbers and atomic numbers on each side of the equation must balance.

Beta particles

$^{0}_{-1}e^{-}$

Beta particles are electrons. They are fast-moving, negatively charged particles that are emitted from the nucleus of an atom.

They have greater penetration than an alpha particle and can pass through air, but they cannot penetrate thin metal foil.

The nuclear equation for beta decay creates a product with an atomic number that has increased, but the mass number is unchanged. For example:

$$^{228}_{88}Ra \rightarrow {}^{228}_{89}Ac + {}^{0}_{-1}e$$

Gamma waves

Gamma radiation is non-particulate. This means that it is not a particle, like alpha or beta radiation, but a form of high energy electromagnetic radiation. The gamma waves are emitted from the nucleus of a radioactive element.

The non-particulate nature of gamma waves means that they do not change the nature of the atom, so there are no nuclear equations for gamma radiation.

Gamma radiation has the greatest penetration of the three types of radioactivity and thick lead or concrete is required to absorb gamma rays.

The penetration of each of the three types of radioactivity is illustrated in Figure 3.25.

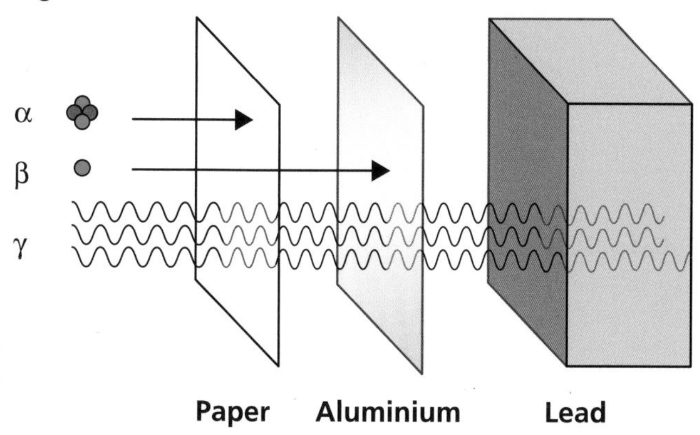

Figure 3.25 The penetration of each of the three types of radioactivity

The nature of each type of radiation can be demonstrated by passing through a magnetic field (Figure 3.26).

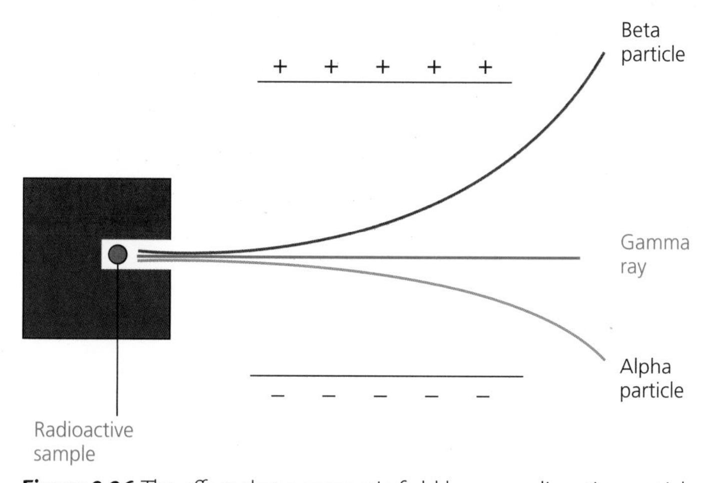

Figure 3.26 The effect that a magnetic field has on radioactive particles

The beta particle $^{0}_{-1}e^{-}$ is negatively charged and, therefore, it is attracted to the positively charged plate.

The alpha particle $^{4}_{2}He^{2+}$ is positively charged and, therefore, it is attracted to the negatively charged plate.

Gamma waves have no charge, because they are non-particulate, so they are not affected by any of the charged plates.

Nuclear equations involving protons ${}^{1}_{1}p$ and neutrons ${}^{1}_{0}n$ can also be written like this:

$${}^{53}_{27}Co \rightarrow {}^{52}_{26}Fe + {}^{1}_{1}P$$

$${}^{13}_{4}Be \rightarrow {}^{12}_{4}Be + {}^{1}_{0}n$$

Half-life

The decay of a radioactive element is constant and unaffected by chemical or physical conditions. This means that the breakdown of unstable nuclei is not affected by temperature, pressure, concentration, mass or chemical state.

This makes it possible to accurately calculate the time taken for the activity or mass of a radioactive element to drop by half. The time it takes to do this is called the **half-life** of the isotope.

Calculating half-life is relatively straightforward as you can see in this example.

Example 

The mass of a radioisotope falls from 3·2 g to 0·2 g in 3 hours. What is the half-life of this radioisotope?

$$\text{original mass} = 3{\cdot}2 \xrightarrow{1} 1{\cdot}6 \xrightarrow{2} 0{\cdot}8 \xrightarrow{3} 0{\cdot}4 \xrightarrow{4} 0{\cdot}2$$

The mass has halved four times in three hours so the half-life is $\frac{3}{4} = 0{\cdot}75$ hours.

Because isotopes decay at a known rate, they can be used to accurately date materials. For example, ${}^{14}_{6}C$ is used to date archaeological specimens up to 10 000 years old. This is illustrated in the half-life graph in Figure 3.27.

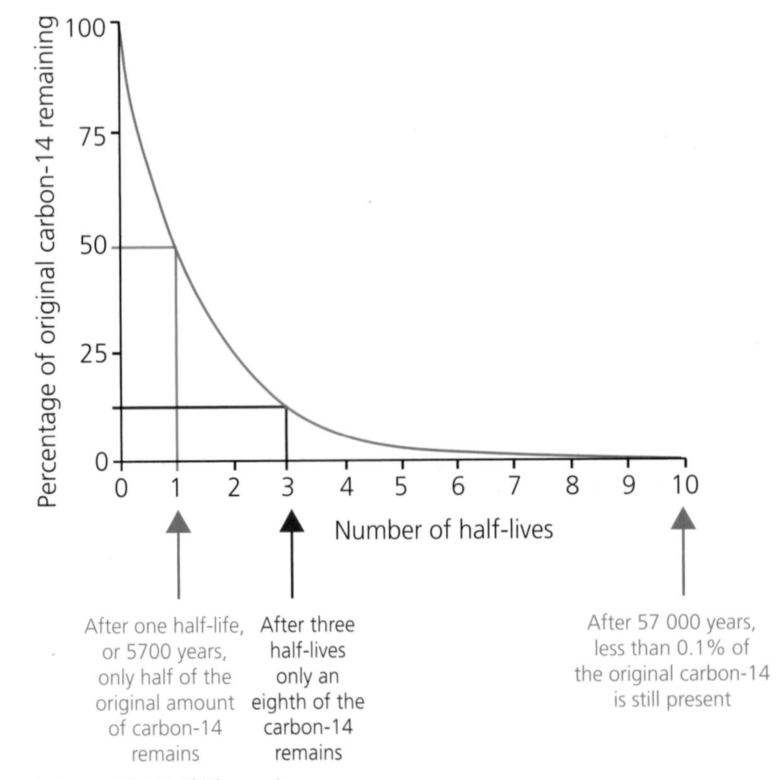

Figure 3.27 Half-life graph

Uses of radioisotopes

The medical and industrial uses of radioisotopes are constantly expanding. We cover a few uses here.

Medical uses

Radioisotopes can be used in the treatment of cancer. Gamma emitter cobalt-60 is used in the treatment of tumours and the less penetrating beta emitter phosphorus-32 can be used in the treatment of skin cancer.

Industrial uses

Leaks in pipelines can be investigated by using radioisotopes with a short half-life. The radiation emitted can be detected if a leak is present.

Americium-241 is used in domestic smoke alarms. Small amounts of smoke can affect the amount of radiation passing through a small gap and an alarm is triggered.

Nuclear power stations that use uranium fuel provide approximately 50% of Scotland's electricity.

Figure 3.28 Torness Nuclear Power Station

Study questions 

1 Radioactive iodine-131 is a very effective treatment of cancer of the thyroid gland. Iodine-131 differs from the stable isotopes of iodine in

a) atomic number
b) atomic mass
c) chemical properties
d) valency.

2 There are three types of radioactive decay, all of which can be stopped in different ways, as shown in Figure 3.29.

Which line in the table correctly identifies the types of radiation identified by particles X, Y and Z?

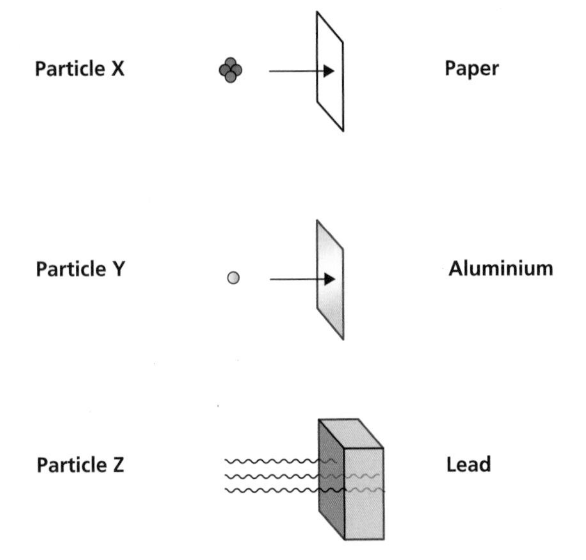

Figure 3.29

	Particle X	Particle Y	Particle Z
A	alpha	beta	gamma
B	beta	alpha	gamma
C	beta	gamma	alpha
D	gamma	beta	alpha

3 An atom of ^{227}Th decays by alpha emission to produce an atom of ^{211}Pb. How many alpha particles were released to produce this lead atom?

a) 1
b) 2
c) 3
d) 4

4 The graph in Figure 3.30 shows the radioactive decay of sodium-24. Choose the correct half-life for this sample from the options below.

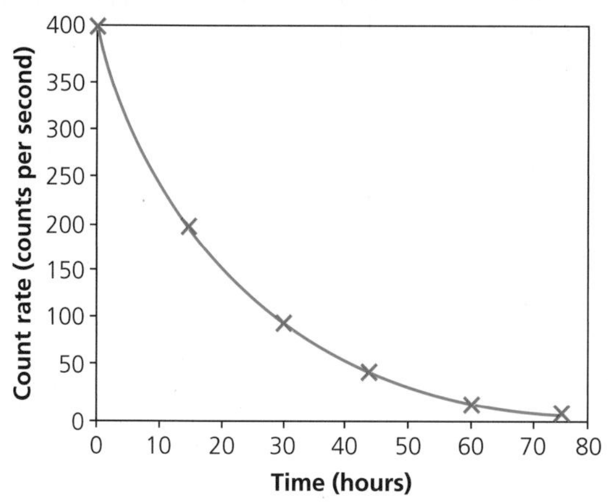

Figure 3.30

a) 10 hours
b) 15 hours
c) 60 hours
d) 200 hours

5 Write a nuclear equation for each of the following:
a) alpha decay of Po-210
b) beta decay of Sr-90
c) alpha decay of Ra-226

6 A 200 g sample of radioactive iodine-131 was weighed accurately. The same sample was reweighed 24 days later and only 25 g of the original sample remained.

Calculate the half-life of iodine-131.

Chapter 3.6 Chemical analysis

It is important that you are familiar with the use of the following pieces of apparatus. Many of them have been included in the diagrams in this book. You must be able to draw clear diagrams of these pieces of apparatus and describe how to use them in experiments.

- conical flask
- beaker
- measuring cylinder
- delivery tube
- dropper
- test tubes/boiling tubes
- funnels
- filter paper
- evaporating basin
- pipette and pipette filler
- burette
- thermometer

You must be aware of the following techniques and be able to describe the procedures. You may be asked to suggest improvements to techniques and procedures to increase the accuracy of the results obtained.

- filtration – see page 45
- using a balance for weighing materials – see page 5
- methods for collection of gases, including:
  - collection over water (for insoluble gases) – see page 4
  - downward displacement of air – (for gases that are less dense than air) – see page 4
  - upward displacement of air (for gases that are more dense than air) – see page 4

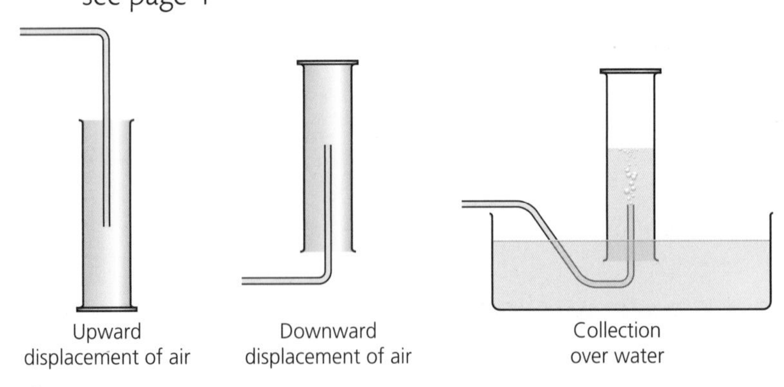

Figure 3.31

- methods of safely heating substances, such as using a heating mantel rather than a Bunsen burner to heat flammable reactants/products
- preparation of soluble salts – see page 45
- preparation of insoluble salts – see page 45
- testing the conductivity of solids and solutions – see page 70
- setting up electrochemical cells – see page 72

- electrolysis, including using a d.c. power supply – see page 76
- determination of E_h – see page 66.

In addition, you must be aware of the following analytical techniques:

- titrations using standard solutions – see page 42
- flame tests to identify solutions of metals salts (the colours produced by different metals are listed on page 6 of the Data Booklet)
- identifying gases such as
  - oxygen – relights a glowing splint (see page 64)
  - carbon dioxide – turns limewater milky/cloudy (see page 64)
  - hydrogen – burns with a pop (see page 69)
- precipitation to form an insoluble salt – (see page 45).

You must be able to communicate the results of experiments by

- drawing accurate, labelled diagrams
- producing line graphs (with lines of best-fit) and bar charts with clear titles, axes labels and units
- producing tables with clear headings that include units
- calculating averages from concordant results (see page 43)
- suggesting improvements to descriptions of experiments to give more accurate results, such as using a heat shield when performing the enthalpy of combustion experiment (see page 65).

Open questions

Open questions usually contain this phrase: '**Using your knowledge of chemistry**'. They provide you with opportunities to demonstrate your detailed knowledge of the subject. The marks are awarded as shown:

- **3 marks** for demonstrating a **good** conceptual understanding of the chemistry involved
- **2 marks** for demonstrating a **reasonable** understanding of the chemistry involved
- **1 mark** for demonstrating a **limited** understanding of the chemistry involved
- **0 marks** for demonstrating no understanding of the chemistry that is relevant to the problem/situation.

When constructing your answer, if possible, you should aim to include:

- a list of bullet points as an introduction or plan, so that – at the very start – you note down the main points that you need to make in your answer
- a diagram of a relevant experiment
- any relevant chemical equations
- other pertinent diagrams to help with your explanation, perhaps a diagram of the electron shells in a question about energy levels for example
- any necessary calculations with workings, such as an average reaction rate for example. Be sure to include units.

An example of an open question and sample answer is given below:

Hydrogen doesn't fit. The first element, hydrogen, has been causing trouble for some time. It can be placed in group 1, as it usually is, or with the halogens in group 7 (Figure 3.32).

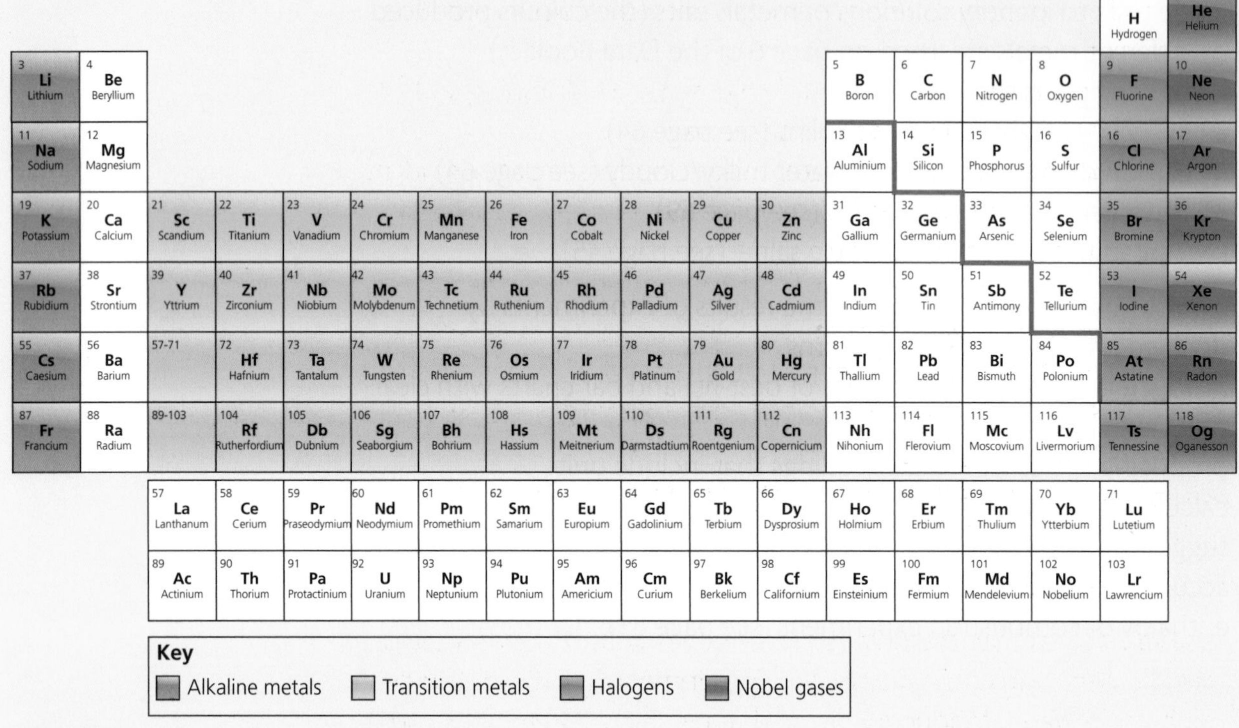

Figure 3.32

Some authors avoid the hydrogen problem altogether by removing it from the main body and by allowing it to float above the rest of the table.

From RSC.org

Using your knowledge of chemistry, give reasons why hydrogen can be placed above group 1 or group 7.

Sample answer

Plan

- Group 1 elements have one outer electron
- Group 7 elements require one electron to become stable
- Group 7 elements are non-metals
- Group 1 elements are metals
- Group 7 elements contain gases.

Hydrogen has one outer electron and so do all the alkali metals. Elements in the same group have the same number of outer electrons, so hydrogen could be placed above group 1. But hydrogen is not a metal.

It could also be placed above group 7 as it requires one electron to achieve a stable electron arrangement (like a noble gas), as do all the elements in group 7.

Fluorine has the electron arrangement of 2, 7 and gains an electron to achieve the same electron arrangement as neon: 2, 8.

$$F + e^- \rightarrow F^-$$

This is a reduction reaction.

Hydrogen could achieve an electron arrangement like the noble gas helium by gaining an electron (reduction). So it could be placed above group 7, although it doesn't have 7 outer electrons.

Group 7 also contains gases and hydrogen is a gas.

This answer would gain three marks because a good knowledge of chemistry has been demonstrated.

Open questions require lots of practice and you should attempt as many as you can as part of your studying. Some example questions are listed below for you to practise. These are given in order of difficulty with the least challenging first.

1. A student is given the task of identifying the type of bonding and elements present in an unknown compound.
   Using your knowledge of chemistry, describe tests that the student could perform to identify both the type of bonding and elements present in the unknown compound.
2. Fuels have developed greatly in the past 200 years.
   Using your knowledge of chemistry, comment on how you could establish the efficiency of a fuel and the types of pollution it creates.
3. Electrolysis of water produces hydrogen and oxygen gases. The hydrogen gas can be used as a fuel.
   Using your knowledge of chemistry, comment on the use of hydrogen as a fuel and its production methods.

Here are some hints to get you started on the questions.

1. First you could cover the three types of bonding: ionic, covalent and metallic. Next you could discuss how properties change with bonding: solubility in water, conduction in solid, liquid and molten states and behaviour during electrolysis, melting and boiling points. You could also look at flame tests.
2. You could write about enthalpy calculations, how to measure energy released ($cm\Delta T$), balanced equations, exothermic reactions, formulae of fuels such as CH_4 and C_2H_5OH.
3. Consider covering electrolysis, ionic solutions, electrolytes, ion-electron equations, the balanced equation for combustion of hydrogen.

Answer section

1.1 Rates of reaction

1 a

2 c

3 a)

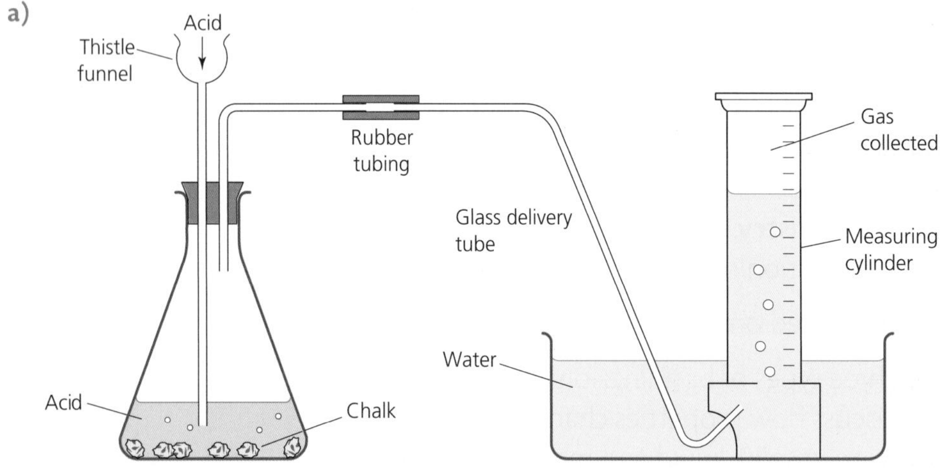

b)

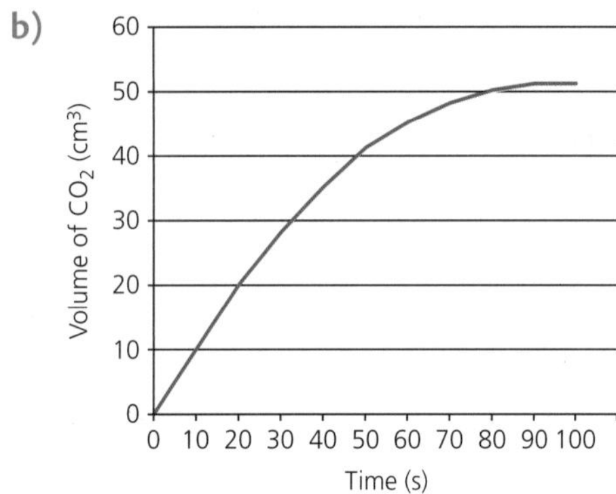

c) $0{\cdot}875\ cm^3\ s^{-1}$

d) The reactants are being used up or the concentration of the reactants decreases as the reaction proceeds.

e) 90 seconds

f)

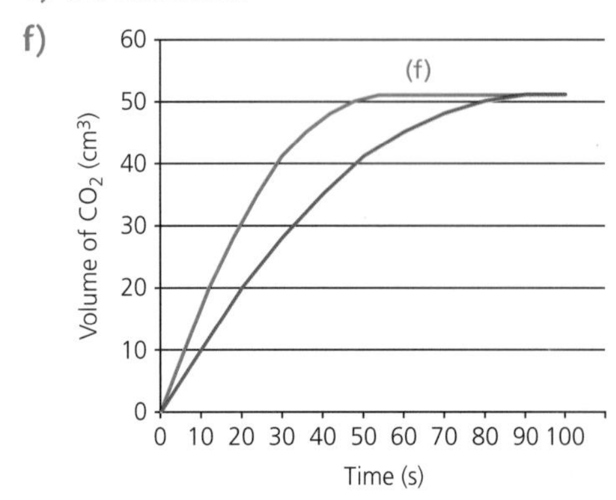

4 Smaller particle size results in a larger surface area, therefore making the catalyst more efficient.

5 a) Liver = 2 $cm^3 s^{-1}$, potato = 0·8 $cm^3 s^{-1}$

b) Liver

c) The volume, temperature and concentration of the reactants. The particle size and mass of the catalyst.

1.2 Atomic theory

1 a

2 a

3 b

4 b

5 B

6 A

7 a)

Particles	Number
Proton	47
Electron	46
Neutron	60

b) Isotopes

c) There is an equal amount of each isotope.

8 a) $^{41}_{20}Ca$

b) 20 protons and 21 neutrons

c) equal numbers of protons and electrons

1.3 + 1.4 Bonding and Properties related to bonding

1 c

2 c

3 c

4 a) A covalent bond is a shared pair of electrons between two non-metal elements. The positive nuclei of each atom have an attraction for the shared electrons creating the bond.

b) i

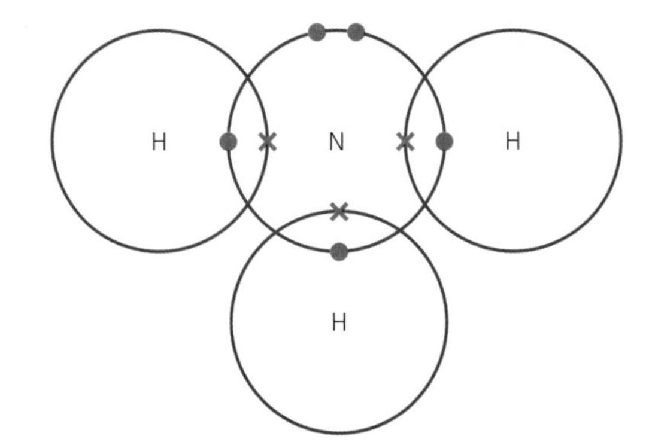

ii

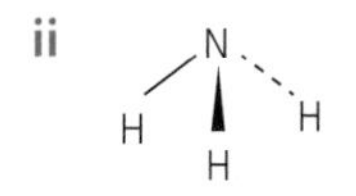

c) i Water is angular; methane is tetrahedral

ii Water, methane and ammonia all have weak intermolecular bonds (attractions) between molecules, which only require low temperatures to break, resulting in low boiling points.

5

Substance	Bonding and structure
D	metallic
C	ionic
B	covalent network
A	covalent molecular

6 a Covalent

b

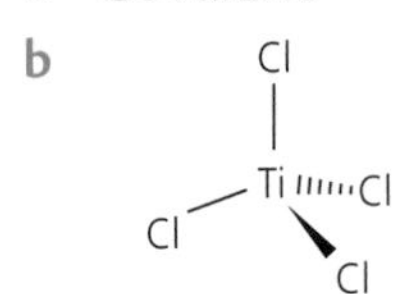

1.5,1.6 + 1.7 Chemical formulae, The mole and Percentage composition calculations

1 a) i MgO

ii $CaCl_2$

iii Al_2S_3

iv NaF

v $Ca(NO_3)_2$

vi $(NH_4)_2SO_4$

b) i $Mg^{2+}\ O^{2-}$

ii $Ca^{2+}\ (Cl^-)_2$

iii $(Al^{3+})_2\ (S^{2-})_3$

iv $Na^+\ F^-$

v $Ca^{2+}\ (NO_3^-)_2$

vi $(NH_4^+)_2\ SO_4^{2-}$

c) i 40·5 g

ii 111 g

iii 150 g

iv 42 g

v 164 g

vi 132 g

2 a) $2Mg + O_2 \rightarrow 2MgO$

b) $CH_4 + 2O_2 \rightarrow CO_2 + 2H_2O$

c) $4Na + O_2 \rightarrow 2Na_2O$

d) $2Li + 2H_2O \rightarrow 2LiOH + H_2$

e) $4Fe + 3O_2 \rightarrow 2Fe_2O_3$

f) $N_2 + 3H_2 \rightarrow 2NH_3$

3 20 g

4 151·5 g

5 2·27 mol l^{-1}

6 a 28·2 %

b 20·8 %

7 88 g

8 52·8 g

1.8 Acids and bases

1 c

2 a

3 b

4 d

5 c

6 b

7 a) sodium oxide + water → sodium hydroxide

b) $Na_2O + H_2O \rightarrow 2NaOH$

c) Above 7

d) Iron oxide is insoluble in water

8 a) 1 mol l^{-1}

b) CH_3COO^- H^+

c) Sodium or ethanoate ion

9 5 cm^3 or 0·005 l or 5×10^{-3} l

2.1 Homologous series

1 b

2 b

3 b

4 c

5 a

6 a) A family of hydrocarbons with similar chemical properties that can be represented by a general formula.

b) CH_2O

7 a) A – remains orange, B – orange to colourless

b)

```
         H   H   H   H
 H       |   |   |   |
  \
   C = C — C — C — C — H
  /
 H           |   |   |
             H   H   H
```

c) Isomers

2.2 Consumer products

1 b

2 c

3 a)

```
      H   H   H
      |   |   |
  H — C — C — C — H
      |   |   |
      H   O   H
          |
          H
```

b) Propan-1-ol

c) 0·01 moles

4 a)

```
       O
      //
H — C
      \
       O — H
```

b) Carboxyl group

c)

```
    H   H   H   H
    |   |   |   |
H — C — C — C — C — O — H
    |   |   |   |
    H   H   H   H
```

2.3 + 2.4 Combustion and Energy from fuels

1 a

2 d

3 a) Exothermic

b) 52·14 kJ

4 80·58 °C

3.1 + 3.2 Metals and Reactions of metals

1 c

2 a

3 a

4 c

5 b

6 b

7 a

8 b

9 a)

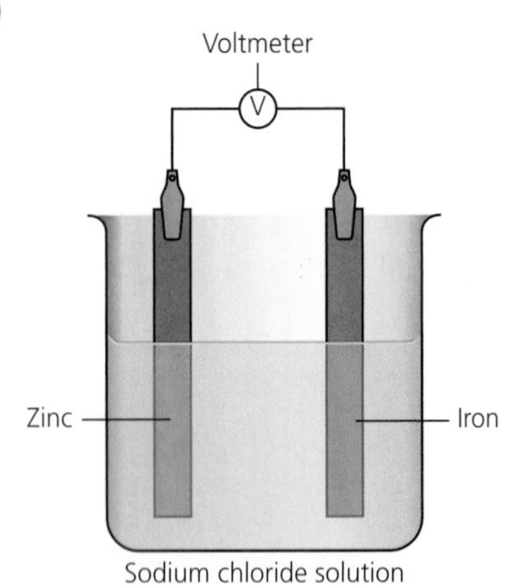

b)

Electrodes	Reading on voltmeter (V)	Direction of electron flow
Zn/Fe	0·3	Zn → Fe
Zn/Al	−0·1	Zn ← Al
Zn/Ni	0·5	Zn → Ni
Zn/Sn	0·7 (value higher than 0.5 V)	Zn → Sn

10 a) To provide oxygen gas

b) Magnesium – Zinc – Copper

11 a) Nitrate

b) $Cu \rightarrow Cu^{2+} + 2e^-$

3.3 + 3.4 Properties of plastics and Fertilisers

1

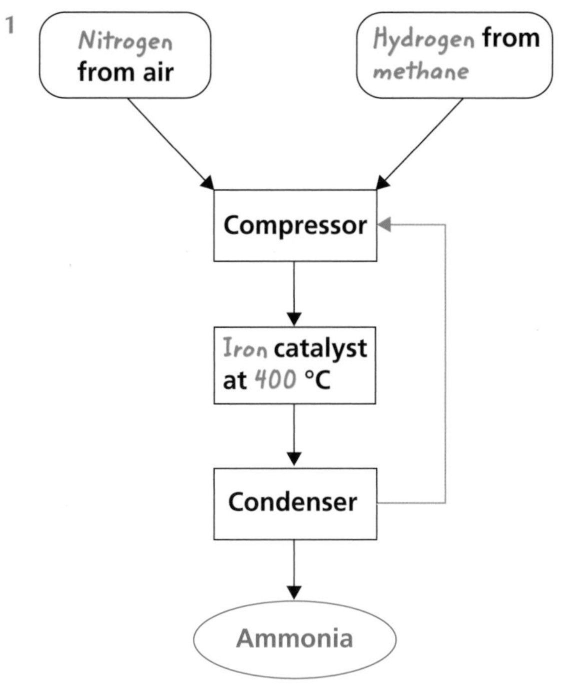

2 a) Addition polymerisation

b)

3 a) 28%

b) 21%

4 a) $NH_4Cl(aq) + NaOH(aq) \rightarrow NH_3(g) + NaCl(aq) + H_2O(l)$

$NH_4^+(aq)\ Cl^-(aq) + Na^+(aq)\ OH^-(aq) \rightarrow NH_3(g) + Na^+(aq)\ Cl^-(aq) + H_2O(l)$

b) Ammonia has a lower density than air.

3.5 Nuclear chemistry

1 b

2 A

3 d

4 b

5 a) $^{210}_{84}Po \rightarrow {}^{206}_{82}Pb + {}^{4}_{2}He^{2+}$

b) $^{90}_{38}Sr \rightarrow {}^{90}_{39}Y + {}^{0}_{-1}e^{-}$

c) $^{226}_{88}Ra \rightarrow {}^{222}_{86}Rn + {}^{4}_{2}He^{2+}$

6 8 days

Key word glossary

Acid (p. 38) A substance with a pH below 7. They contain a higher concentration of H^+ ions than pure water.

Addition polymerisation (p. 79) A reaction in which many small monomers combine to form one large polymer molecule.

Addition reaction (p. 52) A reaction in which a small molecule, usually a diatomic molecule, adds across the double bond of an alkene.

Alcohols (p. 53) A homologous series which contains the –OH functional group and the general formula of $C_nH_{2n+1}OH$.

Alkali (p. 38) A substance with a pH above 7 because it has a higher concentration of OH^- ions than pure water.

Alkali metals (p. 12) The very reactive group 1 metals.

Alkanes (p. 48) The simplest homologous series of saturated hydrocarbons with the general formula C_nH_{2n+2}.

Alkenes (p. 50) A homologous series of hydrocarbons that are unsaturated and have the general formula C_nH_{2n}. They are isomers of the cylcoalkanes.

Alpha particle (p. 86) Slow moving, positively charged particles that come from the nucleus of a radioactive element.

Atom (p. 9) A particle made up of protons, neutrons, and electrons.

Atomic number (p. 10) The number given to each element in the Periodic Table. It is equal to the number of protons in an atom.

Base (p. 39) A substance that neutralises an acid.

Beta particle (p. 86) Fast moving, negatively charged particles.

Carboxyl group (p. 60) The functional group –COOH found in carboxylic acids.

Carboxylic acids (p. 60) A homologous series which contain the functional group –COOH and have the general formula $C_nH_{2n+1}COOH$.

Catalyst (p. 83) Alters the rate of a reaction but is not used up in the reaction.

Compound (p. 19) A substance made up of two or more elements chemically joined.

Covalent bond (p. 17) An attraction between two nucli and a shared pair of electrons in two non-metal atoms.

Covalent molecule (p. 23) Small molecules where the atoms are held together by covalent bonds and only weak attractions form between the molecules.

Covalent network (p. 23) A giant network of atoms held together by covalent bonds. They have very high melting and boiling points.

Cycloalkanes (p. 51) A homologous series with a ring of carbon atoms and saturated bonds. They have the general formula of C_nH_{2n}. They are isomers of the alkenes.

Diatomic compound (p. 20) A compound with molecules containing two atoms only, such as carbon monoxide (CO).

Diatomic element (p. 18) Elements whose molecules contain two atoms, such as chlorine (Cl_2).

Dihaloalkanes (p. 53) The product of an addition reaction between an alkene and a halogen, such as bromine.

Electrochemical cell (p. 70) The connecting of two metals (or a metal and a non-metal) of different reactivity in an electrolyte to produce electricity.

Electrolysis (p. 75) The separation of an ionic solution using a d.c. power supply.

Electrolyte (p. 72) An ionic solution which conducts electricity and completes the circuit.

Electron (p. 9) A negatively charged particle that has a mass of approx 0.

Element (p. 10) A substance that contains only one type of atom.

Endothermic (p. 63) A reaction which takes in energy. This results in a decrease in temperature.

Evaporation (p. 24) Is the process in which a liquid is turned into a gas by heat. It can be used as a separation technique.

Exothermic (p. 63) A reaction that gives out energy. This results in a temperature increase.

Fertiliser (p. 81) A substance that restores elements, essential for healthy plant growth to the soil.

Filtrate (p. 45) The liquid that passes through the filter paper and is collected after filtration.

Filtration (p. 45) A separation technique that separates an insoluble solid from a liquid.

Fuel (p. 51) A substance that reacts exothermically with oxygen.

Functional group (p. 59) The part of a molecule that gives the compound its chemical properties.

Gamma radiation (p. 87) High energy wave emitted from the nucleus of an unstable atom.

Gram formula mass (GFM) (p. 29) The mass of one mole of a substance.

Haber process (p. 82) The industrial method of ammonia production using an iron catalyst.

Half-life (p. 88) The time taken for the activity or mass of a radioactive element to drop by half.

Halogens (p. 53) The reactive non-metals in group 7 of the Periodic Table.

Homologous series (p. 48) A family of compounds with similar chemical properties that can be represented by a general formula.

Hydration (p. 53) The addition of water across a double bond. A type of addition reaction.

Hydrocarbon (p. 48) A compound made up of carbon and hydrogen only.

Hydrogenation (p. 53) The addition of hydrogen across a double bond. A type of addition reaction.

Hydroxyl group (p. 60) The functional group of alcohols (–OH).

Ion (p. 14) A charged particle formed by the losing (metals) or gaining (non-metals) of electrons.

Ion bridge (p. 73) Used to complete the circuit in a cell.

Ionic bond (p. 21) The electrostatic force of attraction between a positive metal ion and a negative non-metal ion.

Ionic lattice (p. 22) A large arrangement of oppositely charged ions held together by ionic bonds. They have high melting and boiling points and dissolve in water. They conduct when molten or in solution.

Isomers (p. 56) Compounds with the same molecular formula but a different structural formula.

Isotope (p. 13) Atoms with the same atomic number but different mass numbers due to a difference in the number of neutrons present in the atoms.

Mass number (p. 12) Equal to the number of protons plus neutrons in an atom.

Metallic bond (p. 68) The electrostatic force of attraction between positively charged ions and delocalised electrons.

Miscible (p. 60) When two substances, such as water and alcohol, are mixed together without separating.

Mole (p. 29) The gram formula mass of a substance.

Molecule (p. 17) Two or more atoms held together by covalent bonds.

Monomer (p. 79) Small unsaturated molecule that combines with other unsaturated molecules to form a polymer.

Neutralisation (p. 39) The reaction of an acid with an alkali or a base that moves the pH towards 7.

Neutron (p. 9) A particle in an atom which is found in the nucleus and has a mass of 1.

Noble gas (p. 14) Very unreactive non-metals found in group 0 of the Periodic Table.

Nucleus (p. 9) The positively charged centre of an atom that contains the neutrons and protons.

Ore (p. 75) A naturally occuring metal compound.

Oxidation reaction (p. 74) A reaction in which electrons are lost.

Percentage composition (p. 35) The percentage, by mass, of an element contained within a compound.

pH (p. 37) A number that indicates the acidity or alkalinity of a substance.

Polymer (p. 79) A very large molecule formed by the addition of many small monomer molecules.

Polymerisation (p. 79) The reaction in which a polymer is formed by the combining of many monomer units.

Precipitate (p. 45) An insoluble solid (salt) produced by a chemical reaction.

Proton (p. 9) A small positively charged particle found in the nucleus of an atom. It has a mass of 1.

Reactivity series (p. 69) A list of metals (and hydrogen) in order of reactivity.

Redox reaction (p. 74) A reaction in which both oxidation and reduction take place.

Reduction reaction (p. 74) A reaction in which electrons are gained.

Relative atomic mass (RAM) (p. 13) The average mass of all the isotopes of an element.

Residue (p. 45) The insoluble solid left over in the filter paper after filtration.

Salt (p. 40) A product of neutralisation in which the hydrogen ion of an acid has been replaced by the ammonium ion or metal ion of the base.

Saturated hydrocarbon (p. 52) A hydrocarbon that has only single carbon-to-carbon bonds. Alkanes and cycloalkanes are saturated.

Solute (p. 30) The substance that is dissolved in the solvent.

Solution (p. 30) The substance produced when a solute dissolves in a solvent.

Solvent (p. 30) A liquid in which a solute dissolves.

Spectator ion (p. 41) Ions that are present during a reaction but remain unchanged by the reaction.

Standard solution (p. 42) A solution of accurately know concentration.

Titration (p. 42) An analytical technique involving accurate measuring of volumes of reacting liquids.

Unsaturated hydrocarbon (p. 52) A hydrocarbon that contains at least one carbon-to-carbon double bond. The alkenes are unsaturated.

Valency (p. 26) The number of bonds that an element or ion can form.